D1121697

DATE DUE

MY 19 '95			
NO 27 '95			
AP 17 '06			

DEMCO 38-296

What Is Philosophy?

European Perspectives

EUROPEAN PERSPECTIVES

European Perspectives

A Series in Social Thought and Cultural Criticism

Lawrence D. Kritzman, Editor

European Perspectives presents English translations of books by leading European thinkers. With both classic and outstanding contemporary works, the series aims to shape the major intellectual controversies of our day and to facilitate the tasks of historical understanding.

Julia Kristeva	*Strangers to Ourselves*
Theodor W. Adorno	*Notes to Literature, vols. 1 and 2*
Richard Wolin, ed.	*The Heidegger Controversy*
Antonio Gramsci	*Prison Notebooks, vol. 1*
Jacques LeGoff	*History and Memory*
Alain Finkielkraut	*Remembering in Vain*
Julia Kristeva	*Nations Without Nationalism*
Pierre Bourdieu	*Field of Cultural Production*
Pierre Vidal-Naquet	*Assassins of Memory*
Hugo Ball	*Critique of the German Intelligentsia*

Gilles Deleuze & Félix Guattari

What Is Philosophy?

Translated by Hugh Tomlinson and Graham Burchell

Columbia University Press   New York

...vishes to express its
appreciation of assistance given by the government
of France through Le Ministère de la Culture in the
preparation of this translation.

COLUMBIA UNIVERSITY PRESS

NEW YORK   CHICHESTER, WEST SUSSEX

Qu'est-ce que la philosophie? © 1991 by Les Editions de
Minuit. Translation © 1994 Columbia University Press

Library of Congress Cataloging-in-Publication Data
Deleuze, Gilles.
    [Qu'est-ce que la philosophie? English]
    What is philosophy? / Gilles Deleuze and Félix Guattari;
translated by Hugh Tomlinson and Graham Burchell.
        p.   cm.—(European perspectives)
    Includes bibliographical references and index.
    ISBN 0–231–07988–5
    1. Philosophy. 2. Science. 3. Logic. 4. Aesthetics.
I. Guattari, Félix. II. Title. III. Series.
B2430.D453Q4713 1994
100–dc20   93–40801
                                                CIP

Casebound editions of
Columbia University Press books
are printed on
permanent and durable
acid-free paper.

Printed in the United States of America
c 10 9 8 7 6 5 4 3 2 1

Contents

For nearly twenty years, the jointly signed works of Gilles Deleuze and Félix Guattari[1] have made an extraordinary impact. This book, which was published in France in 1991, was at the top of the best-seller list for several weeks. But despite its popular success, *What Is philosophy?* is not a primer or a textbook. It more closely resembles a manifesto produced under the slogan "Philosophers of the world, create!" It is a book that speaks about philosophy, and about philosophies and philosophers, but it is even more a book that takes up arms for philosophy. Most of all, perhaps, it is a book of philosophy as a practice of the creation of concepts.[2]

Félix Guattari died on August 29, 1992, at the age of sixty-two. The production of this book

---

1. In order of original publication these are *Anti-Oedipus,* trans. Robert Hurley, Mark Seem, and Helen R. Lane (Minneapolis: University of Minnesota Press, 1983); *Kafka: Toward a Minor Literature,* trans. Dana Polan (Minneapolis: University of Minnesota Press, 1986); *A Thousand Plateaus,* trans. Brian Massumi (Minneapolis: University of Minnesota Press, 1987).

2. For a general discussion of this book see Eric Alliez, *La Signature du Monde: ou, Qu'est-ce que la philosophie de Deleuze et Guattari?* (Paris: Editions du Cerf, 1993).

was therefore the last achievement of a form of experimental "authorship" that has few precedents in philosophy.[3] Deleuze has spoken of their way of working on a number of occasions: "We do not work together, we work between the two. . . . We don't work, we negotiate. We were never in the same rhythm, we were always out of step."[4] The interaction with Guattari the nonphilosopher brought the philosopher Deleuze to a new stage: from thinking the multiple to doing the multiple.

This process of "a parallel evolution" is exemplified in the "conceptual vitalism" of this book. Deleuze and Guattari are the thinkers of "lines of flight," of the openings that allow thought to escape from the constraints that seek to define and enclose creativity. This conception and practice of philosophy as conceptual creation poses some special difficulties for the translator, as

> some concepts must be indicated by an extraordinary and sometimes even barbarous or shocking word, whereas others make do with an ordinary, everyday word that is filled with harmonics so distant that it risks being imperceptible to a nonphilosophical ear. Some concepts call for archaisms, and others for neologisms, shot through with almost crazy etymological exercises.[5]

In translating such words our first aim has been consistency. We have sought to use the same English word on each occasion. Furthermore, we have tried to avoid departure from other recent translations of Deleuze and Guattari's works. The translation of these key terms is marked with translators' notes. We have tried to keep

---

3. Deleuze's own production shows no sign of diminishing after forty years of writing. His latest work, *Critique et Clinique* (Paris: Minuit, 1993) was published on September 8, 1993. He is at present writing a work on "the greatness of Marx."

4. Gilles Deleuze and Claire Parnet, *Dialogues,* trans. Hugh Tomlinson and Barbara Habberjam (Minneapolis: University of Minnesota Press, 1987), p. 17.

5. Ibid., pp. 7–8.

such notes to a minimum; they are indicated by an asterisk and appear at the bottom of the page.

A number of terms used throughout the book present particular difficulties. There are various English translations of *chiffre,* for example. These include "figure," "numeral," "sum total," "initials" or "monogram," "secret code" or "cipher." None of these capture the philosophical use of the word in the present work. In most instances, we have rendered *chiffre* as "combination" to indicate an identifying numeral (in the sense of the combination of a safe or an opus number, as in music) of a multiplicity, but which is not, however, a number in the sense of a measure.

The word *voisinage* here has the general sense of "neighborhood" but also its mathematical sense, as in "neighborhood of a point," which in a linear set (for example, the points of a straight line) is an open segment containing this point. *Ordonnée* can have the general sense of "ordered." Deleuze and Guattari also use the word in the more technical sense of "ordinate" (as in the vertical, or y-coordinate of Cartesian geometry) in contrast with "abscissa" (the horizontal or x-coordinate).

It is difficult to find a single English equivalent for the word *survol*. The word derives from *survoler,* "to fly over" or "to skim or rapidly run one's eyes over something." However, the present use derives from the philosopher Raymond Ruyer.[6] Ruyer uses the notion of an absolute or nondimensional "survol" to describe the relationship of the "I-unity" to the subjective sensation of a visual field. This sensation, he says, tempts us to imagine the "I" as a kind of invisible center outside, and situated in a supplementary dimension perpendicular to, the whole of the visual field that it surveys from a distance. However, this is an error. The immediate survey of the unity of the visual field made up of many different details takes place within the dimension of the visual sensation itself; it is a kind of "self-enjoy-

6. In *Neo-Finalisme* (Paris: PUF, 1952), especially chap. 9.

ment" that does not involve any supplementary dimension. We have therefore rendered *survol* as "survey."[7]

We would like to thank all those who have given us support and assistance, including in particular Martin Joughin and Michèle Le Dœuff. Finally, we would like to thank our editors at Columbia University Press for their assistance and persistence in the face of our continual attempts to deterritorialize their schedules. This translation is dedicated to Georgia and Felix and to Bebb.

Hugh Tomlinson

Graham Burchell

7. See also Gilles Deleuze, *The Logic of Sense,* trans. Mark Lester (New York: Columbia University Press, 1990), in which *survolant* is translated as "surveying."

What Is Philosophy?

Introduction: The Question Then...

The question *what is philosophy?* can perhaps be posed only late in life, with the arrival of old age and the time for speaking concretely. In fact, the bibliography on the nature of philosophy is very limited. It is a question posed in a moment of quiet restlessness, at midnight, when there is no longer anything to ask. It was asked before; it was always being asked, but too indirectly or obliquely; the question was too artificial, too abstract. Instead of being seized by it, those who asked the question set it out and controlled it in passing. They were not sober enough. There was too much desire to *do* philosophy to wonder what it was, except as a stylistic exercise. That point of nonstyle where one can finally say, "What is it I have been doing all my life?" had not been reached. There are times when old age produces not eternal youth but a sovereign freedom, a pure necessity in which one enjoys a moment of grace between life and death, and in which all the parts of the machine come together to send into the future a feature that cuts across all ages: Titian, Turner, Monet.[1] In old age Turner acquired or won the right to take painting down a deserted path of no return that is indistinguishable from a final question. *Vie de Rancé* could be said to mark both Chateaubriand's old age and the start of modern literature.[2] Cinema too sometimes offers us its gifts of the third age, as when Ivens, for example, blends his laughter with the witch's laughter in the howling wind. Likewise in philosophy, Kant's *Critique of Judgment* is an unrestrained work of old age, which his successors have still not caught up with: all the mind's faculties overcome their limits, the very limits that Kant had so carefully laid down in the works of his prime.

We cannot claim such a status. Simply, the time has come for us to ask what philosophy is. We had never stopped asking this question previously, and we already had the answer, which has not changed: philosophy is the art of forming, inventing, and fabricating concepts. But the answer not only had to take note of the question, it had to determine its moment, its occasion and circumstances, its landscapes and personae, its conditions and unknowns. It had to be possible to ask the question "between friends," as a secret or a confidence, or as a challenge when confronting the enemy, and at the same time to reach that twilight hour when one distrusts even the friend. It is then that you say, "That's what it was, but I don't know if I really said it, or if I was convincing enough." And you realize that having said it or been convincing hardly matters because, in any case, that is what it is now.

We will see that concepts need conceptual personae [*personnages conceptuels**] that play a part in their definition. *Friend* is one such persona that is even said to reveal the Greek origin of philo-sophy:

*Deleuze's and Guattari's *personnages conceptuel* has affiliations with Messiaen's *personnages rythmiques,* which Brian Massumi translates as "rhythmic characters" in Gilles Deleuze and Félix Guattari, *A Thousand Plateaus* (London: Athlone, 1988). We have preferred *persona* and *personae* to *character* and *characters* in order to emphasize the distinction between Deleuze's and Guattari's notion and a more general notion other civilizations had sages, but the Greeks introduce these "friends" who are not just more modest sages. The Greeks might seem to have confirmed the death of the sage and to have replaced him with philosophers—the friends of wisdom, those who seek wisdom but do not formally possess it.[3] But the difference between the sage and the philosopher would not be merely one of degree, as on a scale: the old oriental sage thinks, perhaps, in Figures, whereas the philosopher invents and thinks the Concept. Wisdom has changed a great deal. It is even more difficult to know what *friend* signifies, even and especially among the Greeks. Does it designate a type of competent intimacy, a sort of material taste and potentiality, like that of the joiner with wood—is the potential of wood latent in the good joiner; is he the friend of the wood? The question is important because the friend who appears in philosophy no longer stands for an extrinsic persona, an example or empirical circumstance, but rather for a presence that is intrinsic to thought, a condition of possibility of thought itself, a living category, a transcendental lived reality [*un vécu transcendental*]. With the creation of philosophy, the Greeks violently force the friend into a relationship that is no longer a relationship with an other but one with an Entity, an Objectality [*Objectité**], an Essence—Plato's friend, but even more the friend of wisdom, of truth or the concept, like Philalethes and Theophilus. The philosopher is expert in concepts and in the lack of them. He knows which of them are not viable, which are arbitrary or inconsistent, which ones do not hold up for an instant. On the other hand, he also knows which are well formed and attest to a creation, however disturbing or dangerous it may be.

What does *friend* mean when it becomes a conceptual persona, or

---

of *characters* referring to any figures appearing, for example, in a philosophical dialogue.

*In her translation of Sartre's *Being and Nothingness* (New York: Philosophical Library, 1956), Hazel Barnes translates *objectité,* which she glosses as "the quality or state of being an object" (p. 632), as "objectness" or, on occasions, as "object-state."

a condition for the exercise of thought? Or rather, are we not talking of the lover? Does not the friend reintroduce into thought a vital relationship with the Other that was supposed to have been excluded from pure thought? Or again, is it not a matter of someone other than the friend or lover? For if the philosopher is the friend or lover of wisdom, is it not because he lays claim to wisdom, striving for it potentially rather than actually possessing it? Is the friend also the claimant then, and is that of which he claims to be the friend the Thing to which he lays claim but not the third party who, on the contrary, becomes a rival? Friendship would then involve competitive distrust of the rival as much as amorous striving toward the object of desire. The basic point about friendship is that the two friends are like claimant and rival (but who could tell them apart?). It is in this first aspect that philosophy seems to be something Greek and coincides with the contribution of cities: the formation of societies of friends or equals but also the promotion of relationships of rivalry between and within them, the contest between claimants in every sphere, in love, the games, tribunals, the judiciaries, politics, and even in thought, which finds its condition not only in the friend but in the claimant and the rival (the dialectic Plato defined as *amphisbetesis*). It is the rivalry of free men, a generalized athleticism: the agon.[4] Friendship must reconcile the integrity of the essence and the rivalry of claimants. Is this not too great a task?

Friend, lover, claimant and rival are transcendental determinations that do not for that reason lose their intense and animated existence, in one persona or in several. When again today Maurice Blanchot, one of the rare thinkers to consider the meaning of the word *friend* in philosophy, takes up this question internal to the conditions of thought as such, does he not once more introduce new conceptual personae into the heart of the purest Thought? But in this case the

---

We have preferred "objectality," in line with Massumi's translation of *visagéité* as "faciality" in *A Thousand Plateaus*.

personae are hardly Greek, arriving from elsewhere as if they had gone through a catastrophe that draws them toward new living relationships raised to the level of a priori characteristics—a turning away, a certain tiredness, a certain distress between friends that converts friendship itself to thought of the concept as distrust and infinite patience?[5] The list of conceptual personae is never closed and for that reason plays an important role in the evolution or transformations of philosophy. The diversity of conceptual personae must be understood without being reduced to the already complex unity of the Greek philosopher.

The philosopher is the concept's friend; he is potentiality of the concept. That is, philosophy is not a simple art of forming, inventing, or fabricating concepts, because concepts are not necessarily forms, discoveries, or products. More rigorously, philosophy is the discipline that involves *creating* concepts. Does this mean that the friend is friend of his own creations? Or is the actuality of the concept due to the potential of the friend, in the unity of creator and his double? The object of philosophy is to create concepts that are always new. Because the concept must be created, it refers back to the philosopher as the one who has it potentially, or who has its power and competence. It is no objection to say that creation is the prerogative of the sensory and the arts, since art brings spiritual entities into existence while philosophical concepts are also "sensibilia." In fact, sciences, arts, and philosophies are all equally creative, although only philosophy creates concepts in the strict sense. Concepts are not waiting for us ready-made, like heavenly bodies. There is no heaven for concepts. They must be invented, fabricated, or rather created and would be nothing without their creator's signature. Nietzsche laid down the task of philosophy when he wrote, "[Philosophers] must no longer accept concepts as a gift, nor merely purify and polish them, but first *make* and *create* them, present them and make them convincing. Hitherto one has generally trusted one's concepts as if they were a wonderful dowry from some sort of wonderland," but trust must be

replaced by distrust, and philosophers must distrust most those concepts they did not create themselves (Plato was fully aware of this,
even though he taught the opposite).[6] Plato said that Ideas must be
contemplated, but first of all he had to create the concept of Idea.
What would be the value of a philosopher of whom one could say,
"He has created no concepts; he has not created his own concepts"?

We can at least see what philosophy is not: it is not contemplation,
reflection, or communication. This is the case even though it may
sometimes believe it is one or other of these, as a result of the capacity
of every discipline to produce its own illusions and to hide behind its
own peculiar smokescreen. It is not contemplation, for contemplations are things themselves as seen in the creation of their specific
concepts. It is not reflection, because no one needs philosophy to
reflect on anything. It is thought that philosophy is being given a
great deal by being turned into the art of reflection, but actually it
loses everything. Mathematicians, as mathematicians, have never
waited for philosophers before reflecting on mathematics, nor artists
before reflecting on painting or music. So long as their reflection
belongs to their respective creation, it is a bad joke to say that this
makes them philosophers. Nor does philosophy find any final refuge
in communication, which only works under the sway of opinions in
order to create "consensus" and not concepts. The idea of a Western
democratic conversation between friends has never produced a single
concept. The idea comes, perhaps, from the Greeks, but they dis-
trusted it so much, and subjected it to such harsh treatment, that
the concept was more like the ironical soliloquy bird that surveyed
[*survolait*] the battlefield of destroyed rival opinions (the drunken
guests at the banquet). Philosophy does not contemplate, reflect, or
communicate, although it must create concepts for these actions or
passions. Contemplation, reflection and communication are not disciplines but machines for constituting Universals in every discipline.
The Universals of contemplation, and then of reflection, are like two
illusions through which philosophy has already passed in its dream of

dominating the other disciplines (objective idealism and subjective idealism). Moreover, it does no credit to philosophy for it to present itself as a new Athens by falling back on Universals of communication that would provide rules for an imaginary mastery of the markets and the media (intersubjective idealism). Every creation is singular, and the concept as a specifically philosophical creation is always a singularity. The first principle of philosophy is that Universals explain nothing but must themselves be explained.

To know oneself, to learn to think, to act as if nothing were self-evident—wondering, "wondering that there is being"—these, and many other determinations of philosophy create interesting attitudes, however tiresome they may be in the long run, but even from a pedagogical point of view they do not constitute a well-defined occupation or precise activity. On the other hand, the following definition of philosophy can be taken as being decisive: knowledge through pure concepts. But there is no reason to oppose knowledge through concepts and the construction of concepts within possible experience on the one hand and through intuition on the other. For, according to the Nietzschean verdict, you will know nothing through concepts unless you have first created them—that is, constructed them in an intuition specific to them: a field, a plane, and a ground that must not be confused with them but that shelters their seeds and the personae who cultivate them. Constructivism requires every creation to be a construction on a plane that gives it an autonomous existence. To create concepts is, at the very least, to make something. This alters the question of philosophy's use or usefulness, or even of its harmfulness (to whom is it harmful?).

Many problems hurry before the hallucinating eyes of an old man who sees all sorts of philosophical concepts and conceptual personae confronting one another. First, concepts are and remain signed: Aristotle's substance, Descartes's cogito, Leibniz's monad, Kant's condition, Schelling's power, Bergson's duration [durée]. But also, some concepts must be indicated by an extraordinary and sometimes even

barbarous or shocking word, whereas others make do with an ordinary, everyday word that is filled with harmonics so distant that it
risks being imperceptible to a nonphilosophical ear. Some concepts
call for archaisms, and others for neologisms, shot through with
almost crazy etymological exercises: etymology is like a specifically
philosophical athleticism. In each case there must be a strange necessity for these words and for their choice, like an element of style.
The concept's baptism calls for a specifically philosophical *taste* that
proceeds with violence or by insinuation and constitutes a philosophical language within language—not just a vocabulary but a syntax
that attains the sublime or a great beauty. Although concepts are
dated, signed, and baptized, they have their own way of not dying
while remaining subject to constraints of renewal, replacement, and
mutation that give philosophy a history as well as a turbulent geography, each moment and place of which is preserved (but in time) and
that passes (but outside time). What unity remains for philosophies,
it will be asked, if concepts constantly change? Is it the same for the
sciences and arts that do not work with concepts? And what are their
respective histories the histories of? If philosophy is this continuous
creation of concepts, then obviously the question arises not only of
what a concept is as philosophical Idea but also of the nature of the
other creative Ideas that are not concepts and that are due to the arts
and sciences, which have their own history and becoming and which
have their own variable relationships with one another and with
philosophy. The exclusive right of concept creation secures a function
for philosophy, but it does not give it any preeminence or privilege
since there are other ways of thinking and creating, other modes of
ideation that, like scientific thought, do not have to pass through
concepts. We always come back to the question of the use of this
activity of creating concepts, in its difference from scientific or artistic
activity. Why, through what necessity, and for what use must concepts, and always new concepts, be created? And in order to do

what? To say that the greatness of philosophy lies precisely in its not having any use is a frivolous answer that not even young people find amusing any more. In any case, the death of metaphysics or the overcoming of philosophy has never been a problem for us: it is just tiresome, idle chatter. Today it is said that systems are bankrupt, but it is only the concept of system that has changed. So long as there is a time and a place for creating concepts, the operation that undertakes this will always be called philosophy, or will be indistinguishable from philosophy even if it is called something else.

We know, however, that the friend or lover, as claimant, does not lack rivals. If we really want to say that philosophy originates with the Greeks, it is because the city, unlike the empire or state, invents the agon as the rule of a society of "friends," of the community of free men as rivals (citizens). This is the invariable situation described by Plato: if each citizen lays claim to something, then we need to be able to judge the validity of claims. The joiner lays claim to wood, but he comes up against the forester, the lumberjack, and the carpenter, who all say, "I am the friend of wood." If it is a matter of the care of men, then there are many claimants who introduce themselves as man's friend: the peasant who feeds people, the weaver who clothes them, the doctor who nurses them, and the warrior who protects them.[7] In all these cases the selection is made from what is, after all, a somewhat narrow circle of claimants. But this is not the case in politics where, according to Plato, anyone can lay claim to anything in Athenian democracy. Hence the necessity for Plato to put things in order and create authorities for judging the validity of these claims: the Ideas as philosophical concepts. But, even here, do we not encounter all kinds of claimants who say, "I am the true philosopher, the friend of Wisdom or of the Well-Founded"? This rivalry culminates in the battle between philosopher and sophist, fighting over the old sage's remains. How, then, is the false friend to be distinguished from the true friend, the concept from the simulacrum? The simulator and the friend: this is a whole Platonic theater that produces a proliferation of conceptual personae by endowing them with the powers of the comic and the tragic.

Closer to our own time, philosophy has encountered many new rivals. To start with, the human sciences, and especially sociology, wanted to replace it. But because philosophy, taking refuge in universals, increasingly misunderstood its vocation for creating concepts, it was no longer clear what was at stake. Was it a matter of giving up the creation of concepts in favor of a rigorous human science or, alternatively, of transforming the nature of concepts by turning them into the collective representations or worldviews created by the vital, historical, and spiritual forces of different peoples? Then it was the turn of epistemology, of linguistics, or even of psychoanalysis and logical analysis. In successive challenges, philosophy confronted increasingly insolent and calamitous rivals that Plato himself would never have imagined in his most comic moments. Finally, the most shameful moment came when computer science, marketing, design, and advertising, all the disciplines of communication, seized hold of the word *concept* itself and said: "This is our concern, we are the creative ones, we are the *ideas men!* We are the friends of the concept, we put it in our computers." Information and creativity, concept and enterprise: there is already an abundant bibliography. Marketing has preserved the idea of a certain relationship between the concept and the event. But here the concept has become the set of product displays (historical, scientific, artistic, sexual, pragmatic), and the event has become the exhibition that sets up various displays and the "exchange of ideas" it is supposed to promote. The only events are exhibitions, and the only concepts are products that can be sold. Philosophy has not remained unaffected by the general movement that replaced Critique with sales promotion. The simulacrum, the simulation of a packet of noodles, has become the true concept; and the one who packages the product, commodity, or work of art has become the philosopher, conceptual persona, or artist. How could philosophy, an old person, compete against young executives in a race for the universals of communication for determining the marketable form of the concept, *Merz?** Certainly, it is painful to learn that *Concept* indicates a society of information services and engineering. But the more philosophy comes up against shameless and inane rivals and encounters them at its very core, the more it feels driven to fulfill the task of creating concepts that are aerolites rather than commercial products. It gets the giggles, which wipe away its tears. So, the question of philosophy is the singular point where concept and creation are related to each other.

Philosophers have not been sufficiently concerned with the nature of the concept as philosophical reality. They have preferred to think of it as a given knowledge or representation that can be explained by the faculties able to form it (abstraction or generalization) or employ it (judgment). But the concept is not given, it is created; it is to be created. It is not formed but posits itself in itself—it is a self-positing. Creation and self-positing mutually imply each other because what is truly created, from the living being to the work of art, thereby enjoys a self-positing of itself, or an autopoetic characteristic by which it is recognized. The concept posits itself to the same extent that it is created. What depends on a free creative activity is also that which, independently and necessarily, posits itself in itself: the most subjective will be the most objective. The post-Kantians, and notably Schelling and Hegel, are the philosophers who paid most attention to the concept as philosophical reality in this sense. Hegel powerfully defined the concept by the Figures of its creation and the Moments of its self-positing. The figures become parts of the concept because they constitute the aspect through which the concept is created by and in consciousness, through successive minds; whereas the Mo-

---

*Merz is the term coined by the artist Kurt Schwitters to refer to the aesthetic combination of any kind of material, and the equal value of these different materials, in his collages and assemblages. The term itself came from a fragment of a word in one of his assemblages, the whole phrase being "Kommerz und Privatbank."

ments form the other aspect according to which the concept posits
itself and unites minds in the absolute of the Self. In this way Hegel
showed that the concept has nothing whatever to do with a general
or abstract idea, any more than with an uncreated Wisdom that does
not depend on philosophy itself. But he succeeded in doing this at
the cost of an indeterminate extension of philosophy that, because it
reconstituted universals with its own moments and treated the personae of its own creation as no more than ghostly puppets, left scarcely
any independent movement of the arts and sciences remaining. The
post-Kantians concentrated on a universal *encyclopedia* of the concept
that attributed concept creation to a pure subjectivity rather than
taking on the more modest task of a *pedagogy* of the concept, which
would have to analyze the conditions of creation as factors of always
singular moments.[8] If the three ages of the concept are the encyclopedia, pedagogy, and commercial professional training, only the second
can safeguard us from falling from the heights of the first into the
disaster of the third—an absolute disaster for thought whatever its
benefits might be, of course, from the viewpoint of universal capitalism.

Part One

Philosophy

1. What Is a Concept?

There are no simple concepts. Every concept has components and is defined by them. It therefore has a combination [*chiffre**]. It is a multiplicity, although not every multiplicity is conceptual. There is no concept with only one component. Even the first concept, the one with which a philosophy "begins," has several components, because it is not obvious that philosophy must have a beginning, and if it does determine one, it must combine it with a point of view or a ground [*une raison*]. Not only do Descartes, Hegel, and Feuerbach not begin with the same concept, they do not have the same concept of beginning. Every concept is at least double or triple, etc. Neither is there a concept possessing every component, since this would be chaos pure and simple. Even so-called universals as ultimate concepts must escape the chaos by circumscribing a universe that explains them (contemplation, reflection, communication). Every concept has an irregular contour defined by the sum of its compo-

*See translators' introduction.

nents, which is why, from Plato to Bergson, we find the idea of the concept being a matter of articulation, of cutting and cross-cutting. The concept is a whole because it totalizes its components, but it is a fragmentary whole. Only on this condition can it escape the mental chaos constantly threatening it, stalking it, trying to reabsorb it.

On what conditions is a concept first, not absolutely but in relation to another? For example, is *another person* [*autrui*] necessarily second in relation to a self? If so, it is to the extent that its concept is that of an other—a subject that presents itself as an object—which is special in relation to the self: they are two components. In fact, if the other person is identified with a special object, it is now only the other subject as it appears to me; and if we identify it with another subject, it is me who is the other person as I appear to that subject. All concepts are connected to problems without which they would have no meaning and which can themselves only be isolated or understood as their solution emerges. We are dealing here with a problem concerning the plurality of subjects, their relationship, and their reciprocal presentation. Of course, everything changes if we think that we discover another problem: what is the nature of the other person's position that the other subject comes to "occupy" only when it appears to me as a special object, and that I in turn come to occupy as special object when I appear to the other subject? From this point of view the other person is not anyone—neither subject nor object. There are several subjects because there is the other person, not the reverse. The other person thus requires an a priori concept from which the special object, the other subject, and the self must all derive, not the other way around. The order has changed, as has the nature of the concepts and the problems to which they are supposed to respond. We put to one side the question of the difference between scientific and philosophical problems. However, even in philosophy, concepts are only created as a function of problems which are thought to be badly understood or badly posed (pedagogy of the concept).

Let us proceed in a summary fashion: we will consider a field of experience taken as a real world no longer in relation to a self but to a simple "there is." There is, at some moment, a calm and restful world. Suddenly a frightened face looms up that looks at something out of the field. The other person appears here as neither subject nor object but as something that is very different: a possible world, the possibility of a frightening world. This possible world is not real, or not yet, but it exists nonetheless: it is an expressed that exists only in its expression—the face, or an equivalent of the face. To begin with, the other person is this existence of a possible world. And this possible world also has a specific reality in itself, as possible: when the expressing speaks and says, "I am frightened," even if its words are untruthful, this is enough for a reality to be given to the possible as such. This is the only meaning of the "I" as linguistic index. But it is not indispensable: China is a possible world, but it takes on a reality as soon as Chinese is spoken or China is spoken about within a given field of experience. This is very different from the situation in which China is realized by becoming the field of experience itself. Here, then, is a concept of the other that presupposes no more than the determination of a sensory world as condition. On this condition the other appears as the expression of a possible. The other is a possible world as it exists in a face that expresses it and takes shape in a language that gives it a reality. In this sense it is a concept with three inseparable components: possible world, existing face, and real language or speech.

Obviously, every concept has a history. This concept of the other person goes back to Leibniz, to his possible worlds and to the monad as expression of the world. But it is not the same problem, because in Leibniz possibles do not exist in the real world. It is also found in the modal logic of propositions. But these do not confer on possible worlds the reality that corresponds to their truth conditions (even when Wittgenstein envisages propositions of fear or pain, he does not see them as modalities that can be expressed in a position of the other person because he leaves the other person oscillating between another subject and a special object). Possible worlds have a long history.[1] In short, we say that every concept always has a *history,* even though this history zigzags, though it passes, if need be, through other problems or onto different planes. In any concept there are usually bits or components that come from other concepts, which corresponded to other problems and presupposed other planes. This is inevitable because each concept carries out a new cutting-out, takes on new contours, and must be reactivated or recut.

On the other hand, a concept also has a *becoming* that involves its relationship with concepts situated on the same plane. Here concepts link up with each other, support one another, coordinate their contours, articulate their respective problems, and belong to the same philosophy, even if they have different histories. In fact, having a finite number of components, every concept will branch off toward other concepts that are differently composed but that constitute other regions of the same plane, answer to problems that can be connected to each other, and participate in a co-creation. A concept requires not only a problem through which it recasts or replaces earlier concepts but a junction of problems where it combines with other coexisting concepts. The concept of the Other Person as expression of a possible world in a perceptual field leads us to consider the components of this field for itself in a new way. No longer being either subject of the field or object in the field, the other person will become the condition under which not only subject and object are redistributed but also figure and ground, margins and center, moving object and reference point, transitive and substantial, length and depth. The Other Person is always perceived as an other, but in its concept it is the condition of all perception, for others as for ourselves. It is the condition for our passing from one world to another. The Other Person makes the world go by, and the "I" now designates only a past world ("I was peaceful"). For example, the Other Person is enough to make any length a possible depth in space, and vice versa, so that if this concept did not function in the perceptual field, transitions and inversions would become incomprehensible, and we would always run up against things, the possible having disappeared. Or at least, philosophically, it would be necessary to find another reason for not running up against them. It is in this way that, on a determinable plane, we go from one concept to another by a kind of bridge. The creation of a concept of the Other Person with these components will entail the creation of a new concept of perceptual space, with other components to be determined (not running up against things, or not too much, will be part of these components).

We started with a fairly complex example. How could we do otherwise, because there is no simple concept? Readers may start from whatever example they like. We believe that they will reach the same conclusion about the nature of the concept or the concept of *concept*. First, every concept relates back to other concepts, not only in its history but in its becoming or its present connections. Every concept has components that may, in turn, be grasped as concepts (so that the Other Person has the face among its components, but the Face will itself be considered as a concept with its own components). Concepts, therefore, extend to infinity and, being created, are never created from nothing. Second, what is distinctive about the concept is that it renders components inseparable *within itself*. Components, or what defines the *consistency* of the concept, its endoconsistency, are distinct, heterogeneous, and yet not separable. The point is that each partially overlaps, has a zone of neighborhood [*zone de voisinage**], or a threshold of indiscernibility, with another one. For example, in the concept of the other person, the possible world does not exist outside the face that expresses it, although it is distinguished from it as expressed and expression; and the face in turn is the vicinity of the words for which it is already the megaphone. Components remain distinct, but something passes from one to the other, some-

*See translator's introduction

thing that is undecidable between them. There is an area *ab* that belongs to both *a* and *b,* where *a* and *b* "become" indiscernible. These zones, thresholds, or becomings, this inseparability, define the internal consistency of the concept. But the concept also has an exoconsistency with other concepts, when their respective creation implies the construction of a bridge on the same plane. Zones and bridges are the joints of the concept.

Third, each concept will therefore be considered as the point of coincidence, condensation, or accumulation of its own components. The conceptual point constantly traverses its components, rising and falling within them. In this sense, each component is an *intensive feature,* an intensive ordinate [*ordonnée intensive**], which must be understood not as general or particular but as a pure and simple singularity—"a" possible world, "a" face, "some" words—that is particularized or generalized depending upon whether it is given variable values or a constant function. But, unlike the position in science, there is neither constant nor variable in the concept, and we no more pick out a variable species for a constant genus than we do a constant species for variable individuals. In the concept there are only ordinate relationships, not relationships of comprehension or extension, and the concept's components are neither constants nor variables but pure and simple *variations* ordered according to their neighborhood. They are processual, modular. The concept of a bird is found not in its genus or species but in the composition of its postures, colors, and songs: something indiscernible that is not so much synesthetic as syneidetic. A concept is a heterogenesis—that is to say, an ordering of its components by zones of neighborhood. It is ordinal, an intension present in all the features that make it up. The concept is in a state of *survey* [*survol†*] in relation to its components, endlessly traversing them according to an order without distance. It

*See translators' introduction.
†See translators' introduction.

is immediately co-present to all its components or variations, at no distance from them, passing back and forth through them: it is a refrain, an opus with its number (*chiffre*).

The concept is an incorporeal, even though it is incarnated or effectuated in bodies. But, in fact, it is not mixed up with the state of affairs in which it is effectuated. It does not have spatiotemporal coordinates, only intensive ordinates. It has no energy, only intensities; it is anenergetic (energy is not intensity but rather the way in which the latter is deployed and nullified in an extensive state of affairs). The concept speaks the event, not the essence or the thing— pure Event, a hecceity, an entity: the event of the Other or of the face (when, in turn, the face is taken as concept). It is like the bird as event. The concept is defined by *the inseparability of a finite number of heterogeneous components traversed by a point of absolute survey at infinite speed*. Concepts are "absolute surfaces or volumes," forms whose only object is the inseparability of distinct variations.[2] The "survey" [*survol*] is the state of the concept or its specific infinity, although the infinities may be larger or smaller according to the number of components, thresholds and bridges. In this sense the concept is act of thought, it is thought operating at infinite (although greater or lesser) speed.

The concept is therefore both absolute and relative: it is relative to its own components, to other concepts, to the plane on which it is defined, and to the problems it is supposed to resolve; but it is absolute through the condensation it carries out, the site it occupies on the plane, and the conditions it assigns to the problem. As whole it is absolute, but insofar as it is fragmentary it is relative. It is *infinite through its survey or its speed but finite through its movement that traces the contour of its components*. Philosophers are always recasting and even changing their concepts: sometimes the development of a point of detail that produces a new condensation, that adds or withdraws components, is enough. Philosophers sometimes exhibit a forgetfulness that almost makes them ill. According to Jaspers, Nietzsche, "corrected his ideas himself in order to create new ones without explicitly admitting it; when his health deteriorated he forgot the conclusions he had arrived at earlier." Or, as Leibniz said, "I thought I had reached port; but . . . I seemed to be cast back again into the open sea."[3] What remains absolute, however, is the way in which the created concept is posited in itself and with others. The relativity and absoluteness of the concept are like its pedagogy and its ontology, its creation and its self-positing, its ideality and its reality—the concept is real without being actual, ideal without being abstract. The concept is defined by its consistency, its endoconsistency and exoconsistency, but it has no *reference:* it is self-referential; it posits itself and its object at the same time as it is created. Constructivism unites the relative and the absolute.

Finally, the concept is not discursive, and philosophy is not a discursive formation, because it does not link propositions together. Confusing concept and proposition produces a belief in the existence of scientific concepts and a view of the proposition as a genuine "intension" (what the sentence expresses). Consequently, the philosophical concept usually appears only as a proposition deprived of sense. This confusion reigns in logic and explains its infantile idea of philosophy. Concepts are measured against a "philosophical" grammar that replaces them with propositions extracted from the sentences in which they appear. We are constantly trapped between alternative propositions and do not see that the concept has already passed into the excluded middle. The concept is not a proposition at all; it is not propositional, and the proposition is never an intension. Propositions are defined by their reference, which concerns not the Event but rather a relationship with a state of affairs or body and with the conditions of this relationship. Far from constituting an intension, these conditions are entirely extensional. They imply operations by which abscissas or successive linearizations are formed that force intensive ordinates into spatiotemporal and energetic coordinates, by which the sets so determined are made to correspond to each other.

These successions and correspondences define discursiveness in extensive systems. The *independence of variables* in propositions is opposed to the *inseparability of variations* in the concept. Concepts, which have only consistency or intensive ordinates outside of any coordinates, freely enter into relationships of nondiscursive resonance—either because the components of one become concepts with other heterogeneous components or because there is no difference of scale between them at any level. Concepts are centers of vibrations, each in itself and every one in relation to all the others. This is why they all resonate rather than cohere or correspond with each other. There is no reason why concepts should cohere. As fragmentary totalities, concepts are not even the pieces of a puzzle, for their irregular contours do not correspond to each other. They do form a wall, but it is a dry-stone wall, and everything holds together only along diverging lines. Even bridges from one concept to another are still junctions, or detours, which do not define any discursive whole. They are movable bridges. From this point of view, philosophy can be seen as being in a perpetual state of digression or digressiveness.

The major differences between the philosophical enunciation of fragmentary concepts and the scientific enunciation of partial propositions follow from this digression. From an initial point of view, all enunciation is positional. But enunciation remains external to the proposition because the latter's object is a state of affairs as referent, and the references that constitute truth values as its conditions (even if, for their part, these conditions are internal to the object). On the other hand, positional enunciation is strictly immanent to the concept because the latter's sole object is the inseparability of the components that constitute its consistency and through which it passes back and forth. As for the other aspect, creative or signed enunciation, it is clear that scientific propositions and their correlates are just as signed or created as philosophical concepts: we speak of Pythagoras's theorem, Cartesian coordinates, Hamiltonian number, and Lagrangian function just as we speak of the Platonic Idea or Descartes's cogito and the like. But however much the use of proper names clarifies and confirms the historical nature of their link to these enunciations, these proper names are masks for other becomings and serve only as pseudonyms for more secret singular entities. In the case of propositions, proper names designate extrinsic *partial observers* that are scientifically definable in relation to a particular axis of reference; whereas for concepts, proper names are intrinsic *conceptual personae* who haunt a particular plane of consistency. It is not only proper names that are used very differently in philosophies, sciences, and arts but also syntactical elements, and especially prepositions and the conjunctions, "now," "therefore." Philosophy proceeds by sentences, but it is not always propositions that are extracted from sentences in general. At present we are relying only on a very general hypothesis: from sentences or their equivalent, philosophy extracts *concepts* (which must not be confused with general or abstract ideas), whereas science extracts *prospects* (propositions that must not be confused with judgments), and art extracts *percepts and affects* (which must not be confused with perceptions or feelings). In each case language is tested and used in incomparable ways—but in ways that do not define the difference between disciplines without also constituting their perpetual interbreeding.

EXAMPLE I

To start with, the preceding analysis must be confirmed by taking the example of one of the best-known signed philosophical concepts, that of the Cartesian cogito, Descartes's *I:* a concept of self. This concept has three components—doubting, thinking, and being (although this does not mean that every concept must be triple). The complete statement of the concept qua multiplicity is "I think 'therefore' I am" or, more completely, "Myself who doubts, I think, I am, I am a thinking thing." According to Descartes the cogito is the always-renewed event of thought.

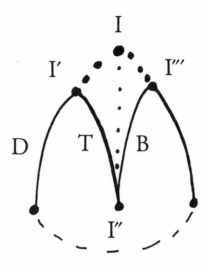

The concept condenses at the point I, which passes through all the components and in which I' (doubting), I" (thinking), and I''' (being) coincide. As intensive ordinates the components are arranged in zones of neighborhood or indiscernibility that produce passages from one to the other and constitute their inseparability. The first zone is between doubting and thinking (myself who doubts, I cannot doubt that I think), and the second is between thinking and being (in order to think it is necessary to be). The components are presented here as verbs, but this is not a rule. It is sufficient that there are variations. In fact, doubt includes moments that are not the species of a genus but the *phases* of a variation: perceptual, scientific, obsessional doubt (every concept therefore has a phase space, although not in the same way as in science). The same goes for modes of thought—feeling, imagining, having ideas—and also for types of being, thing, or substance—infinite being, finite thinking being, extended being. It is noteworthy that in the last case the concept of self retains only the second phase of being and excludes the rest of the variation. But this is precisely the sign that the concept *is closed* as fragmentary totality with "I am a thinking thing": we can pass to other phases of being only by bridges or crossroads that lead to other concepts. Thus, "among my ideas I have the idea of infinity" is the bridge leading from the concept of self to the concept of God. This new concept has three components forming the "proofs" of the existence of God as infinite event. The third (ontological proof) assures the closure of the concept but also in turn throws out a bridge or branches off to a concept of the extended, insofar as it guarantees the objective truth value of our other clear and distinct ideas.

When the question "Are there precursors of the cogito?" is asked, what is meant is "Are there concepts signed by previous philosophers that have similar or almost identical components but from which one component is lacking, or to which others have been added, so that a cogito does not crystallize since the components do not yet coincide in a self?" Everything seems ready, and yet something is missing. Perhaps the earlier concept referred to a different problem from that of the cogito (a change in problems being necessary for the Cartesian cogito to appear), or it was developed on another plane. The Cartesian plane consists in challenging any explicit objective presupposition where every concept refers to other concepts (the rational-animal man, for example). It demands only a prephilosophical understanding, that is, implicit and subjective presuppositions: everyone knows what thinking, being, and I mean (one knows by doing it, being it, or saying it). This is a very novel distinction. Such a plane requires a first concept that presupposes nothing objective. So the problem is "What is the first concept on this plane, or by beginning with what concept can truth as absolutely pure subjective certainty be determined?" Such is the cogito. The other concepts will be able to achieve objectivity, but only if they are linked by bridges to the first concept, if they respond to problems subject to the same conditions, and if they remain on the same plane. Objectivity here will assume a certainty of knowledge rather than presuppose a truth recognized as preexisting, or already there.

There is no point in wondering whether Descartes was right or wrong. Are implicit and subjective presuppositions more valid than explicit objective presuppositions? Is it necessary "to begin," and, if so, is it necessary to start from the point of view of a subjective certainty? Can thought as such be the verb of an I? There is no direct answer. Cartesian concepts can only be assessed as a function of their problems and their plane. In general, if earlier concepts were able to prepare a concept but not constitute it, it is because their problem was still trapped within other problems, and their plane did not yet possess its indispensable curvature or movements. And concepts can only be replaced by others if there are new problems and another plane relative to which (for example) "I" loses all meaning, the beginning loses all necessity, and the presuppositions lose all difference—or take on others. A concept always has the truth that falls to it as a function of the conditions of its creation. Is there one plane that is better than all the others, or problems that dominate all others? Nothing at all can be said on this point. Planes must be constructed and problems posed, just as concepts must be created. Philosophers do the best they can, but they have too much to do to know whether it is the best, or even to bother with this question. Of course, new concepts must relate to our problems, to our history, and, above all, to our becomings. But what does it mean for a concept to be of our time, or of any time? Concepts are not eternal, but does this mean they are temporal? What is the philosophical form of the problems of a particular time? If one concept is "better" than an earlier one, it is because it makes us aware of new variations and unknown resonances, it carries out unforeseen cuttings-out, it brings forth an Event that surveys [*survole*] us. But did the earlier concept not do this already? If one can still be a Platonist, Cartesian, or Kantian today, it is because one is justified in thinking that their concepts can be reactivated in our problems and inspire those concepts that need to be created. What is the best way to follow the great philosophers? Is it to repeat what they said or *to do what they did,* that is, create concepts for problems that necessarily change?

For this reason philosophers have very little time for discussion. Every philosopher runs away when he or she hears someone say, "Let's discuss this." Discussions are fine for roundtable talks, but philosophy throws its numbered dice on another table. The best one can say about discussions is that they take things no farther, since the participants never talk about the same thing. Of what concern is it to philosophy that someone has such a view, and thinks this or that, if the problems at stake are not stated? And when they are stated, it is no longer a matter of discussing but rather one of creating concepts for the undiscussible problem posed. Communication always comes too early or too late, and when it comes to creating, conversation is always superfluous. Sometimes philosophy is turned into the idea of a perpetual discussion, as "communicative rationality," or as "universal democratic conversation." Nothing is less exact, and when philosophers criticize each other it is on the basis of problems and on a plane that is different from theirs and that melt down the old concepts in the way a cannon can be melted down to make new weapons. It never takes place on the same plane. To criticize is only to establish that a concept vanishes when it is thrust into a new milieu, losing some of its components, or acquiring others that transform it. But those who criticize without creating, those who are content to defend the vanished concept without being able to give it the forces it needs to return to life, are the plague of philosophy. All these debaters and communicators are inspired by *ressentiment*. They speak only of themselves when they set empty generalizations against one another. Philosophy has a horror of discussions. It always has something else to do. Debate is unbearable to it, but not because it is too sure of itself. On the contrary, it is its uncertainties that take it down other, more solitary paths. But in Socrates was philosophy not a free discussion among friends? Is it not, as the conversation of free men, the summit of Greek sociability? In fact, Socrates constantly made all discussion impossible, both in the short form of the contest of questions and answers and in the long form of a rivalry between discourses. He turned the friend into the friend of the single concept, and the concept into the pitiless monologue that eliminates the rivals one by one.

EXAMPLE 2

The *Parmenides* shows the extent to which Plato is master of the concept. The One has two components (being and nonbeing), phases of components (the One superior to being, equal to being, inferior to being; the One superior to nonbeing, equal to nonbeing), and zones of indiscernibility (in relation to itself, in relation to others). It is a model concept.

But is not the One prior to every concept? This is where Plato teaches the opposite of what he does: he creates concepts but needs to set them up as representing the uncreated that precedes them. He puts time into the concept, but it is a time that must be Anterior. He constructs the concept but as something that attests to the preexistence of an objectality [*objectité*], in the form of a difference of time capable of measuring the distance or closeness of the concept's possible constructor. Thus, on the Platonic plane, truth is posed as presupposition, as already there. This is the Idea. In the Platonic concept of the Idea, *first* takes on a precise sense, very different from the meaning it will have in Descartes: it is that which objectively possesses a pure quality, or which is not something other than what it is. Only Justice is just, only Courage courageous, such are Ideas, and there is an Idea of mother if there is a mother who is not something other than a mother (who would not have been a daughter), or of hair which is not something other than hair (not silicon as well). Things, on the contrary, are understood as always being something other than what they are. At best, therefore, they only possess quality in a secondary way, they can only *lay claim* to quality, and only to the degree that they *participate* in the Idea. Thus the concept of Idea has the following components: the quality possessed or to be possessed; the Idea that possesses it first, as unparticipable; that which lays claim to the quality and can only possess it second, third, fourth; and the Idea participated in, which judges the claims—the Father, a double of the father, the daughter and the suitors, we might say. These are the intensive ordinates of the Idea: a claim will be justified only through a neighborhood, a greater or lesser proximity it "has had" in relation to the Idea, in the survey of an always necessarily anterior time. *Time in this form of anteriority* belongs to the concept; it is like its zone. Certainly, the cogito cannot germinate on this Greek plane, this Platonic soil. So long as the preexistence of the Idea remains (even in the Christian form of archetypes in God's understanding), the cogito could be prepared but not fully accomplished. For Descartes to create this concept, the meaning of "first" must undergo a remarkable change, take on a subjective meaning; and all difference of time between the idea and the soul that forms it as subject must be annulled (hence the importance of Descartes's point against reminiscence, in which he says that innate ideas do not exist "before" but "at the same time" as the soul). It will be necessary to arrive at an instantaneity of the concept and for God to create even truths. The claim must change qualitatively: the suitor no longer receives the daughter from the father but owes her hand only to his own chivalric prowess—to his own method. Whether Malebranche can reactivate Platonic components on an authentically Cartesian plane, and at what cost, should be analyzed from this point of view. But we only wanted to show that a concept always has components that can prevent the appearance of another concept or, on the contrary, that can themselves appear only at the cost of the disappearance of other concepts. However, a concept is never valued by reference to what it prevents: it is valued for its incomparable position and its own creation.

Suppose a component is added to a concept: the concept will probably break up or undergo a complete change involving, perhaps, another plane—at any rate, other problems. This is what happens with the Kantian cogito. No doubt Kant constructs a "transcendental" plane that renders doubt useless and changes the nature of the presuppositions once again. But it is by virtue of this very plane that he can declare that if the "I think" is a *determination* that, as such, implies an *undetermined* existence ("I am"), we still do not know how this undetermined comes to be *determinable* and hence in what form it appears as *determined*. Kant therefore "criticizes" Descartes for having said, "I am a thinking substance," because nothing warrants such a claim of the "I." Kant demands the introduction of a new component into the cogito, the one Descartes repressed—time. For it is only in time that my undetermined existence is determinable. But I am only determined in time as a passive and phenomenal self, an always affectable, modifiable, and variable self. The cogito now presents four components: I think, and as such I am active; I have an existence; this existence is only determinable in time as a passive self; I am therefore determined as a passive self that necessarily represents its own thinking activity to itself as an Other (*Autre*) that affects it. This is not another subject but rather the subject who becomes an other. Is this the path of a conversion of the self to the other person? A preparation for "I is an other"? A new syntax, with other ordinates, with other zones of indiscernibility, secured first by the schema and then by the affection of self by self [*soi par soi*], makes the "I" and the "Self" *inseparable*.

The fact that Kant "criticizes" Descartes means only that he sets up a plane and constructs a problem that could not be occupied or completed by the Cartesian cogito. Descartes created the cogito as concept, but by expelling time as *form of anteriority,* so as to make it a simple mode of succession referring to continuous creation. Kant reintroduces time into the cogito, but it is a completely different time from that of Platonic anteriority. This is the creation of a concept. He makes time a component of a new cogito, but on condition of providing in turn a new concept of time: time becomes *form of interiority* with three components—succession, but also simultaneity and permanence. This again implies a new concept of space that can no longer be defined by simple simultaneity and becomes form of exteriority. Space, time, and "I think" are three original concepts linked by bridges that are also junctions—a blast of original concepts. The history of philosophy means that we evaluate not only the historical novelty of the concepts created by a philosopher but also the power of their becoming when they pass into one another.

The same pedagogical status of the concept can be found everywhere: a *multiplicity,* an absolute surface or volume, self-referents, made up of a certain number of inseparable intensive variations according to an order of neighborhood, and traversed by a point in a state of survey. The concept is the contour, the configuration, the constellation of an event to come. Concepts in this sense belong to philosophy by right, because it is philosophy that creates them and never stops creating them. The concept is obviously knowledge—but knowledge of itself, and what it knows is the pure event, which must not be confused with the state of affairs in which it is embodied. The task of philosophy when it creates concepts, entities, is always to extract an event from things and beings, to set up the new event from things and beings, always to give them a new event: space, time, matter, thought, the possible as events.

It is pointless to say that there are concepts in science. Even when science is concerned with the same "objects" it is not from the viewpoint of the concept; it is not by creating concepts. It might be said that this is just a matter of words, but it is rare for words not to involve intentions and ruses. It would be a mere matter of words if it was decided to reserve the concept for science, even if this meant finding another word to designate the business of philosophy. But usually things are done differently. The power of the concept is attributed to science, the concept being defined by the creative methods of science and measured against science. The issue is then whether there remains a possibility of philosophy forming secondary concepts that make up for their own insufficiency by a vague appeal to the "lived." Thus Gilles-Gaston Granger begins by defining the concept as a scientific proposition or function and then concedes that there may, nonetheless, be philosophical concepts that replace reference to the object by correlation to a "totality of the lived" [*totalité du vécu*].[4] But actually, either philosophy completely ignores the concept, or else it enjoys it by right and at first hand, so that there is nothing of it left for science—which, moreover, has no need of the concept and concerns itself only with states of affairs and their conditions. Science needs only propositions or functions, whereas philosophy, for its part, does not need to invoke a lived that would give only a ghostly and extrinsic life to secondary, bloodless concepts. The philosophical concept does not refer to the lived, by way of compensation, but consists, through its own creation, in setting up an event that surveys the whole of the lived no less than every state of affairs. Every concept shapes and reshapes the event in its own way. The greatness of a philosophy is measured by the nature of the events to which its concepts summon us or that it enables us to release in concepts. So the unique, exclusive bond between concepts and philosophy as a creative discipline must be tested in its finest details. The concept belongs to philosophy and only to philosophy.

2. The Plane of Immanence

Philosophical concepts are fragmentary wholes that are not aligned with one another so that they fit together, because their edges do not match up. They are not pieces of a jigsaw puzzle but rather the outcome of throws of the dice. They resonate nonetheless, and the philosophy that creates them always introduces a powerful Whole that, while remaining open, is not fragmented: an unlimited One-All, an "Omnitudo" that includes all the concepts on one and the same plane. It is a table, a plateau, or a slice; it is a plane of consistency or, more accurately, the plane of immanence of concepts, the planomenon. Concepts and plane are strictly correlative, but nevertheless the two should not be confused. The plane of immanence is neither a concept nor the concept of all concepts. If one were to be confused with the other there would be nothing to stop concepts from forming a single one or becoming universals and losing their singularity, and the plane would also lose its openness. Philosophy is a constructivism, and constructivism has two qualitatively different complementary aspects: the creation of concepts and the laying out of a plane. Concepts are like multiple waves, rising and falling, but the plane of immanence is the single wave that rolls them up and unrolls them. The plane envelops infinite movements that pass back and forth through it, but concepts are the infinite speeds of finite movements that, in each case, pass only through their own components. From Epicurus to Spinoza (the incredible book 5) and from Spinoza to Michaux the problem of thought is infinite speed. But this speed requires a milieu that moves infinitely in itself—the plane, the void, the horizon. Both elasticity of the concept and fluidity of the milieu are needed.[1] Both are needed to make up "the slow beings" that we are.

Concepts are the archipelago or skeletal frame, a spinal column rather than a skull, whereas the plane is the breath that suffuses the separate parts. Concepts are absolute surfaces or volumes, formless and fragmentary, whereas the plane is the formless, unlimited absolute, neither surface nor volume but always fractal. Concepts are concrete assemblages, like the configurations of a machine, but the plane is the abstract machine of which these assemblages are the working parts. Concepts are events, but the plane is the horizon of events, the reservoir or reserve of purely conceptual events: not the relative horizon that functions as a limit, which changes with an observer and encloses observable states of affairs, but the absolute horizon, independent of any observer, which makes the event as concept independent of a visible state of affairs in which it is brought about.[2] Concepts pave, occupy, or populate the plane bit by bit, whereas the plane itself is the indivisible milieu in which concepts are distributed without breaking up its continuity or integrity: they occupy it without measuring it out (the concept's combination is not a number) or are distributed without splitting it up. The plane is like a desert that concepts populate without dividing up. The only regions of the plane are concepts themselves, but the plane is all that holds them together. The plane has no other regions than the tribes populating and moving around on it. It is the plane that secures conceptual linkages with ever increasing connections, and it is concepts that secure the populating of the plane on an always renewed and variable curve.

The plane of immanence is not a concept that is or can be thought but rather the image of thought, the image thought gives itself of what it means to think, to make use of thought, to find one's bearings in thought. It is not a method, since every method is concerned with concepts and presupposes such an image. Neither is it a state of knowledge on the brain and its functioning, since thought here is not related to the slow brain as to the scientifically determinable state of affairs in which, whatever its use and orientation, thought is only brought about. Nor is it opinions held about thought, about its forms, ends, and means, at a particular moment. The image of thought implies a strict division between fact and *right:* what pertains to thought as such must be distinguished from contingent features of the brain or historical opinions. *Quid juris?*—can, for example, losing one's memory or being mad belong to thought as such, or are they only contingent features of the brain that should be considered as simple facts? Are contemplating, reflecting, or communicating anything more than opinions held about thought at a particular time and in a particular civilization? The image of thought retains only what thought can claim by right. Thought demands "only" movement that can be carried to infinity. What thought claims by right, what it selects, is infinite movement or the movement of the infinite. It is this that constitutes the image of thought.

Movement of the infinite does not refer to spatiotemporal coordinates that define the successive positions of a moving object and the fixed reference points in relation to which these positions vary. "To orientate oneself in thought" implies neither objective reference point nor moving object that experiences itself as a subject and that, as such, strives for or needs the infinite. Movement takes in everything, and there is no place for a subject and an object that can only be concepts. It is the horizon itself that is in movement: the relative horizon recedes when the subject advances, but on the plane of immanence we are always and already on the absolute horizon. Infinite movement is defined by a coming and going, because it does not advance toward a destination without already turning back on itself, the needle also being the pole. If "turning toward" is the movement of thought toward truth, how could truth not also turn toward thought? And how could truth itself not turn away from thought when thought turns away from it? However, this is not a fusion but a reversibility, an immediate, perpetual, instantaneous exchange—a lightning flash. Infinite movement is double, and there is only a fold from one to the other. It is in this sense that thinking and being are said to be one and the same. Or rather, movement is not the image of thought without being also the substance of being. When Thales's thought leaps out, it comes back as water. When Heraclitus's thought becomes *polemos,* it is fire that retorts. It is a single speed on both sides: "The atom will traverse space with the speed of thought."³ The plane of immanence has two facets as Thought and as Nature, as *Nous* and as *Physis.* That is why there are always many infinite movements caught within each other, each folded in the others, so that the return of one instantaneously relaunches another in such a way that the plane of immanence is ceaselessly being woven, like a gigantic shuttle. To turn toward does not imply merely to turn away but to confront, to lose one's way, to move aside.⁴ Even the negative produces infinite movements: falling into error as much as avoiding the false, allowing oneself to be dominated by passions as much as overcoming them. Diverse movements of the infinite are so mixed in with each other that, far from breaking up the One-All of the plane of immanence, they constitute its variable curvature, its concavities and convexities, its fractal nature as it were. It is this fractal nature that makes the planomenon an infinite that is always different from any surface or volume determinable as a concept. Every movement passes through the whole of the plane by immediately turning back on and folding itself and also by folding other movements or allowing itself to be folded by them, giving rise to retroactions, connections, and proliferations in the fractalization of this infinitely folded up infinity (variable curvature of the plane). But if it is true that the plane of immanence is always single, being itself pure variation, then it is all the more necessary to explain why there are varied and distinct planes of immanence that, depending upon which infinite movements are retained and selected, succeed and contest each other in history. The plane is certainly not the same in the time of the Greeks, in the seventeenth century, and today (and these are still vague and general terms): there is neither the same image of thought nor the same substance of being. The plane is, therefore, the object of an infinite specification so that it seems to be a One-All only in cases specified by the selection of movement. This difficulty concerning the ultimate nature of the plane of immanence can only be resolved step by step.

It is essential not to confuse the plane of immanence and the concepts that occupy it. Although the same elements may appear twice over, on the plane and in the concept, it will not be in the same guise, even when they are expressed in the same verbs and words. We have seen this for being, thought, and one: they enter into the concept's components and are themselves concepts, but they belong to the plane quite differently as image or substance. Conversely, truth can only be defined on the plane by a "turning toward" or by "that toward which thought turns"; but this does not provide us with a concept of truth. If error itself is an element that by right forms part of the plane, then it consists simply in taking the false for the true (falling); but it only receives a concept if we determine its components (according to Descartes, for example, the two components of a finite understanding and an infinite will). Movements or elements of the plane, therefore, will seem to be only nominal definitions in relation to concepts so long as we disregard the difference in nature between plane and concepts. But in reality, elements of the plane are *diagrammatic features,* whereas concepts are *intensive features*. The former are movements of the infinite, whereas the latter are intensive ordinates of these movements, like original sections or differential positions: finite movements in which the infinite is now only speed and each of which constitutes a surface or a volume, an irregular contour marking a halt in the degree of proliferation. The former are *directions* that are fractal in nature, whereas the latter are absolute *dimensions,* intensively defined, always fragmentary surfaces or volumes. The former are *intuitions,* and the latter *intensions.* The grandiose Leibnizian or Bergsonian perspective that every philosophy depends upon an intuition that its concepts constantly develop through slight differences of intensity is justified if intuition is thought of as the envelopment of infinite movements of thought that constantly pass through a plane of immanence. Of course, we should not conclude from this that concepts are deduced from the plane: concepts require a special construction distinct from that of the plane, which is why concepts must be created just as the plane must be set up. Intensive features are never the consequence of diagrammatic features, and intensive ordinates are not deduced from movements or directions. Their correspondence goes beyond even simple resonances and introduces instances adjunct to the creation of concepts, namely, conceptual personae.

If philosophy begins with the creation of concepts, then the plane of immanence must be regarded as prephilosophical. It is presupposed not in the way that one concept may refer to others but in the way that concepts themselves refer to a nonconceptual understanding. Once again, this intuitive understanding varies according to the way in which the plane is laid out. In Descartes it is a matter of a subjective understanding implicitly presupposed by the "I think" as first concept; in Plato it is the virtual image of an already-thought that doubles every actual concept. Heidegger invokes a "preontological understanding of Being," a "preconceptual" understanding that seems to imply the grasp of a substance of being in relationship with a predisposition of thought. In any event, philosophy posits as prephilosophical, or even as nonphilosophical, the power of a One-All like a moving desert that concepts come to populate. Prephilosophical does not mean something preexistent but rather something *that does not exist outside philosophy,* although philosophy presupposes it. These are its internal conditions. The nonphilosophical is perhaps closer to the heart of philosophy than philosophy itself, and this means that philosophy cannot be content to be understood only philosophically or conceptually, but is addressed essentially to non-philosophers as well.[5] We will see that this constant relationship with nonphilosophy has various features. According to this first feature, philosophy defined as the creation of concepts implies a distinct but inseparable presupposition. Philosophy is at once concept creation and instituting of the plane. The concept is the beginning of philosophy, but the plane is its instituting.[6] The plane is clearly not a program, design, end, or means: it is a plane of immanence that constitutes the absolute ground of philosophy, its earth or deterritorialization, the foundation on which it creates its concepts. Both the creation of concepts and the instituting of the plane are required, like two wings or fins.

Thinking provokes general indifference. It is a dangerous exercise nevertheless. Indeed, it is only when the dangers become obvious that indifference ceases, but they often remain hidden and barely perceptible, inherent in the enterprise. Precisely because the plane of immanence is prephilosophical and does not immediately take effect with concepts, it implies a sort of groping experimentation and its layout resorts to measures that are not very respectable, rational, or reasonable. These measures belong to the order of dreams, of pathological processes, esoteric experiences, drunkenness, and excess. We head for the horizon, on the plane of immanence, and we return with bloodshot eyes, yet they are the eyes of the mind. Even Descartes had his dream. To think is always to follow the witch's flight. Take Michaux's plane of immanence, for example, with its infinite, wild movements and speeds. Usually these measures do not appear in the result, which must be grasped solely in itself and calmly. But then "danger" takes on another meaning: it becomes a case of obvious consequences when pure immanence provokes a strong, instinctive disapproval in public opinion, and the nature of the created concepts strengthens this disapproval. This is because one does not think without becoming something else, something that does not think—an animal, a molecule, a particle—and that comes back to thought and revives it.

The plane of immanence is like a section of chaos and acts like a sieve. In fact, chaos is characterized less by the absence of determinations than by the infinite speed with which they take shape and vanish. This is not a movement from one determination to the other but, on the contrary, the impossibility of a connection between them, since one does not appear without the other having already disappeared, and one appears as disappearance when the other disappears as outline. Chaos is not an inert or stationary state, nor is it a chance mixture. Chaos makes chaotic and undoes every consistency in the infinite. The problem of philosophy is to acquire a consistency without losing the infinite into which thought plunges (in this respect chaos has as much a mental as a physical existence). *To give consistency without losing anything of the infinite* is very different from the problem of science, which seeks to provide chaos with reference points, on condition of renouncing infinite movements and speeds and of carrying out a limitation of speed first of all. Light, or the relative horizon, is primary in science. Philosophy, on the other hand, proceeds by presupposing or by instituting the plane of immanence: it is the plane's variable *curves* that retain the infinite movements that turn back on themselves in incessant exchange, but which also continually free other movements which are retained. The concepts can then mark out the intensive ordinates of these infinite movements, as movements which are themselves finite which form, at infinite speed, variable *contours* inscribed on the plane. By making a section of chaos, the plane of immanence requires a creation of concepts.

To the question "Can or must philosophy be regarded as Greek?" a first answer seemed to be that the Greek city actually appears as the new society of "friends," with all the ambiguities of that word. Jean-Pierre Vernant adds a second answer: the Greeks were the first to conceive of a strict immanence of Order to a cosmic milieu that sections chaos in the form of a plane. If we call such a plane-sieve Logos, the logos is far from being like simple "reason" (as when one says the world is rational). Reason is only a concept, and a very impoverished concept for defining the plane and the movements that pass through it. In short, the first philosophers are those who institute a plane of immanence like a sieve stretched over the chaos. In this sense they contrast with sages, who are religious personae, priests, because they conceive of the institution of an always transcendent order imposed from outside by a great despot or by one god higher than the others, inspired by Eris,* pursuing wars that go beyond any agon and hatreds that object in advance to the trials of rivalry.[7] Whenever there is transcendence, vertical Being, imperial State in the sky or on earth, there is religion; and there is Philosophy whenever there is immanence, even if it functions as arena for the agon and rivalry (the Greek tyrants do not constitute an objection to this, because they are wholeheartedly on the side of the society of friends such as it appears in their wildest, most violent rivalries). Perhaps these two possible determinations of philosophy as Greek are profoundly linked. Only friends can set out a plane of immanence as a ground from which idols have been cleared. In Empedocles, Love lays out the plane, even if she does not return to the self without enfolding Hatred as movement that has become negative showing a subtranscendence of chaos (the volcano) and a supertranscendence of a god. It may be that the first philosophers still look like priests, or even kings. They borrow the sage's mask—and, as Nietzsche says,

*Eris is the Greek divinity of discord, conflict, and strife, the complementary opposite of Philia, the divinity of union and friendship. See Jean-Pierre Vernant, *The Origins of Greek Thought* (New York: Cornell University Press, 1982), pp. 45–47.

how could philosophy not disguise itself in its early stages? Will it ever stop having to disguise itself? If the instituting of philosophy merges with the presupposition of a prephilosophical plane, how could philosophy not profit from this by donning a mask? It remains the case that the first philosophers lay out a plane through which unlimited movements pass continually on two sides, one determinable as Physis inasmuch as it endows Being with a substance, and the other as Nous inasmuch as it gives an image to thought. It is Anaximander who distinguishes between the two sides most rigorously by combining the movement of qualities with the power of an absolute horizon, the Apeiron or the Boundless, but always on the same plane. Philosophers carry out a vast diversion of wisdom; they place it at the service of pure immanence. They replace genealogy with a geology.

EXAMPLE 3

Can the entire history of philosophy be presented from the viewpoint of the instituting of a plane of immanence? Physicalists, who insist on the substance of Being, would then be distinguished from noologists, who insist on the image of thought. But a risk of confusion soon arises: rather than this substance of Being or this image of thought being constituted by the plane of immanence itself, immanence will be related to something like a "dative," Matter or Mind. This becomes clear with Plato and his successors. Instead of the plane of immanence constituting the One-All, immanence is immanent "to" the One, so that another One, this time transcendent, is superimposed on the one in which immanence is extended or to which it is attributed: the neo-Platonists' formula will always be a One beyond the One. Whenever immanence is interpreted as immanent "to" something a confusion of plane and concept results, so that the concept becomes a transcendent universal and the plane becomes an attribute in the concept. When misunderstood in this way, the plane of immanence revives the transcendent again: it is a simple field of phenomena that now only possesses in a secondary way that which first of all is attributed to the transcendent unity.

It gets worse with Christian philosophy. The positing of immanence remains pure philosophical instituting, but at the same time it is tolerated only in very small doses; it is strictly controlled and enframed by the demands of an emanative and, above all, creative transcendence. Putting their work and sometimes their lives at risk, all philosophers must prove that the dose of immanence they inject into world and mind does not compromise the transcendence of a God to which immanence must be attributed only secondarily (Nicholas of Cusa, Eckhart, Bruno). Religious authority wants immanence to be tolerated only locally or at an intermediary level, a little like a terraced fountain where water can briefly immanate on each level but on condition that it comes from a higher source and falls lower down (transascendence and transdescendence, as Wahl said). Immanence can be said to be the burning issue of all philosophy because it takes on all the dangers that philosophy must confront, all the condemnations, persecutions, and repudiations that it undergoes. This at least persuades us that the problem of immanence is not abstract or merely theoretical. It is not immediately clear why immanence is so dangerous, but it is. It engulfs sages and gods. What singles out the philosopher is the part played by immanence or fire. Immanence is immanent only to itself and consequently captures everything, absorbs All-One, and leaves nothing remaining to which it could be immanent. In any case, whenever immanence is interpreted as immanent *to* Something, we can be sure that this Something reintroduces the transcendent.

Beginning with Descartes, and then with Kant and Husserl, the cogito makes it possible to treat the plane of immanence as a field of consciousness. Immanence is supposed to be immanent to a pure consciousness, to a thinking subject. Kant will call this subject transcendental rather than transcendent, precisely because it is the subject of the field of immanence of all possible experience from which nothing, the external as well as the internal, escapes. Kant objects to any transcendent use of the synthesis, but he ascribes immanence to the subject of the synthesis as new, subjective unity. He may even allow himself the luxury of denouncing transcendent Ideas, so as to make them the "horizon" of the field immanent to the subject.[8] But, in so doing, Kant discovers the modern way of saving transcendence: this is no longer the transcendence of a Something, or of a One higher than everything (contemplation), but that of a Subject to which the field of immanence is only attributed by belonging to a self that necessarily represents such a subject to itself (reflection). The Greek world that belonged to no one increasingly becomes the property of a Christian consciousness.

Yet one more step: when immanence becomes immanent "to" a transcendental subjectivity, it is at the heart of its own field that the hallmark or figure [*chiffre*] of a transcendence must appear as action now referring to another self, to another consciousness (communication). This is what happens in Husserl and many of his successors who discover in the Other or in the Flesh, the mole of the transcendent within immanence itself. Husserl conceives of immanence as that of the flux lived by subjectivity. But since all this pure and even untamed lived does not *belong* completely to the self that represents it to itself, something transcendent is reestablished on the horizon, in the regions of nonbelonging: first, in the form of an "immanent or primordial transcendence" of a world populated by intentional objects; second, as the privileged transcendence of an intersubjective world populated by other selves; and third, as objective transcendence of an ideal world populated by cultural formations and the human community. In this modern moment we are no longer satisfied with thinking immanence as immanent to a transcendent; *we want to think transcendence within the immanent, and it is from immanence that a breach is expected.* Thus, in Jaspers, the plane of immanence is given the most profound determination as "Encompassing" [*Englobant*], but this encompassing is no more than a reservoir for eruptions of transcendence. The Judeo-Christian word replaces the Greek logos: no longer satisfied with ascribing immanence to something, immanence itself is made to disgorge the transcendent everywhere. No longer content with handing over immanence to the transcendent, we want it to discharge it, reproduce it, and fabricate it itself. In fact this is not difficult—all that is necessary *is for movement to be stopped.*[9] Transcendence enters as soon as movement of the infinite is stopped. It takes advantage of the interruption to reemerge, revive, and spring forth again. The three sorts of Universals—contemplation, reflection, and communication—are like three philosophical eras—Eidetic, Critical, and Phenomenological—inseparable from the long history of an illusion. The reversal of values had to go so far—making us think that immanence is a prison (solipsism) from which the Transcendent will save us.

Sartre's presupposition of an impersonal transcendental field restores the rights of immanence.[10] When immanence is no longer immanent to something other than itself it is possible to speak of a plane of immanence. Such a plane is, perhaps, a radical empiricism: it does not present a flux of the lived that is immanent to a subject and individualized in that which belongs to a self. It presents only events, that is, possible worlds as concepts, and other people as expressions of possible worlds or conceptual personae. The event does not relate the lived to a transcendent subject = Self but, on the contrary, is related to the immanent survey of a field without subject; the Other Person does not restore transcendence to an other self but returns every other self to the immanence of the field surveyed. Empiricism knows only events and other people and is therefore a great creator of concepts. Its force begins from the moment it defines the subject: a *habitus,* a habit, nothing but a habit in a field of immanence, the habit of saying I.

Spinoza was the philosopher who knew full well that immanence was only immanent to itself and therefore that it was a plane traversed by movements of the infinite, filled with intensive ordinates. He is therefore the prince of philosophers. Perhaps he is the only philosopher never to have compromised with transcendence and to have hunted it down everywhere. In the last book of the *Ethics* he produced the movement of the infinite and gave infinite speeds to thought in the third kind of knowledge. There he attains incredible speeds, with such lightning compressions that one can only speak of music, of tornadoes, of wind and strings. He discovered that freedom exists only within immanence. He fulfilled philosophy because he satisfied its prephilosophical presupposition. Immanence does not refer back to the Spinozist substance and modes but, on the contrary, the Spinozist concepts of substance and modes refer back to the plane of immanence as their presupposition. This plane presents two sides to us, extension and thought, or rather its two powers, power of being and power of thinking. Spinoza is the vertigo of immanence from which so many philosophers try in vain to escape. Will we ever be mature enough for a Spinozist inspiration? It happened once with Bergson:

the beginning of *Matter and Memory* marks out a plane that slices through the chaos—both the infinite movement of a substance that continually propagates itself, and the image of thought that everywhere continually spreads a pure consciousness by right (immanence is not immanent "to" consciousness but the other way around).

The plane is surrounded by illusions. These are not abstract misinterpretations or just external pressures but rather thought's mirages. Can they be explained by the sluggishness of our brain, by the ready-made facilitating paths [*frayage*] of dominant opinions, and by our not being able to tolerate infinite movements or master the infinite speeds that crush us (so that we have to stop the movement and make ourselves prisoners of the relative horizon once more)? Yet it is we ourselves who approach the plane of immanence, who are on the absolute horizon. It is indeed necessary, in part at least, that illusions arise from the plane itself, like vapors from a pond, like pre-Socratic exhalations given off by transformations of the elements that are always at work on the plane. Artaud said that "the plane of consciousness" or limitless plane of immanence—what the Indians called *Ciguri*—also engenders hallucinations, erroneous perceptions, bad feelings.[11] We must draw up a list of these illusions and take their measure, just as Nietzsche, following Spinoza, listed the "four great errors." But the list is infinite. First of all there is the *illusion of transcendence,* which, perhaps, comes before all the others (in its double aspect of making immanence immanent to something and of rediscovering a transcendence within immanence itself); then the *illusion of universals* when concepts are confused with the plane. But this confusion arises as soon as immanence is posited as being immanent to something, since this something is necessarily a concept. We think the universal explains, whereas it is what must be explained, and we fall into a triple illusion—one of contemplation or reflection or communication. Then there is the *illusion of the eternal* when it is forgotten that concepts must be created, and then the *illusion of discursiveness* when propositions are confused with concepts. It would be wrong to think that all these illusions logically entail one another like propositions, but they resonate or reverberate and form a thick fog around the plane.

From chaos the plane of immanence takes the determinations with which it makes its infinite movements or its diagrammatic features. Consequently, we can and must presuppose a multiplicity of planes, since no one plane could encompass all of chaos without collapsing back into it; and each retains only movements which can be folded together. The history of philosophy exhibits so many quite distinct planes not just as a result of illusions, of the variety of illusions, and not merely because each plane has its own, constantly renewed, way of restoring transcendence. More profoundly, it is because each plane has its own way of constructing immanence. Each plane carries out a selection of that which is due to thought by right, but this selection varies from one plane to another. Every plane of immanence is a One-All: it is not partial like a scientific system, or fragmentary like concepts, but distributive—it is an "each." *The* plane of immanence is *interleaved*. When comparing particular cases it is no doubt difficult to judge whether there is a single plane or several different ones: do the pre-Socratics have the same image of thought, despite the differences between Heraclitus and Parmenides? Can we speak of a plane of immanence or image of so-called classical thought that continues from Plato to Descartes? It is not just the planes that vary but the way in which they are distributed. Are there more-or-less close or distant points of view that would make it possible to group different layers over a fairly long period or, on the contrary, to separate layers on what seemed to be a common plane? Where, apart from the absolute horizon, would these points of view come from? Can we be satisfied here with a historicism, or with a generalized relativism? In all these respects, the question of the one or the multiple once again becomes the most important one, introducing itself into the plane.

In the end, does not every great philosopher lay out a new plane of immanence, introduce a new substance of being and draw up a new image of thought, so that there could not be two great philosophers on the same plane? It is true that we cannot imagine a great philosopher of whom it could not be said that he has changed what it means to think; he has "thought differently" (as Foucault put it). When we find several philosophies in the same author, is it not because they have changed plane and once more found a new image? We cannot be unaware of Biran's complaint when he was near to death: "I feel a little too old to start the construction again."[12] On the other hand, those who do not renew the image of thought are not philosophers but functionaries who, enjoying a ready-made thought, are not even conscious of the problem and are unaware even of the efforts of those they claim to take as their models. But how, then, can we proceed in philosophy if there are all these layers that sometimes knit together and sometimes separate? Are we not condemned to attempt to lay out our own plane, without knowing which planes it will cut across? Is this not to reconstitute a sort of chaos? That is why every plane is not only interleaved but holed, letting through the fogs that surround it, and in which the philosopher who laid it out is in danger of being the first to lose himself. That so many fogs arise is explained in two ways. Firstly, because thought cannot stop itself from interpreting immanence as immanent to something, the great Object of contemplation, the Subject of reflection, or the Other subject of communication: then transcendence is inevitably reintroduced. And if this cannot be avoided it is because it seems that each plane of immanence can only claim to be unique, to be *the* plane, by reconstituting the chaos it had to ward off: the choice is between transcendence and chaos.

EXAMPLE 4

When the plane selects what is by right due to thought, in order to make its features, intuitions, directions, or diagrammatic movements, it relegates other determinations to the status of mere facts, characteristics of states of affairs, or lived contents. And, of course, philosophy will be able to draw out concepts from these states of affairs inasmuch as it extracts the event from them. That which belongs to thought by right, that which is retained as diagrammatic feature in itself, represses other rival determinations (even if these latter are called upon to receive a concept). Thus Descartes makes error the feature or direction that expresses what is in principle negative in thought. He was not the first to do this, and "error" might be seen as one of the principal features of the classical image of thought. We know that there are many other things in this image that threaten thinking: stupidity, forgetfulness, aphasia, delirium, madness; but all these determinations will be considered as facts that in principle have only a single effect immanent in thought—error, always error. Error is the infinite movement that gathers together the whole of the negative. Can this feature be traced back to Socrates, for whom the person who is wicked (in fact) is someone who is by right "mistaken"? But, if it is true that the *Thaetetus* is a foundation of error, does not Plato hold in reserve the rights of other rival determinations, like the delirium of the *Phaedrus,* so that it seems to us that the image of thought in Plato plots many other tracks?

A major change occurs, not only in concepts but in the image of thought, when ignorance and superstition replace error and prejudice in expressing what by right is the negative of thought: Fontenelle plays a major role here, and what changes at the same time is the infinite movements in which thought is lost and gained. There is an even greater change when Kant shows that thought is threatened less by error than by inevitable illusions that come from within reason, as if from an internal arctic zone where the needle of every compass goes mad. A reorientation of the whole of thought becomes necessary at the same time as it is in principle penetrated by a certain delirium. It is no longer threatened on the plane of immanence by the holes or ruts of a path that it follows but by Nordic fogs that cover everything. The meaning of the question of "finding one's bearings in thought" itself changes.

A feature cannot be isolated. In fact, the movement given a negative sign is itself folded within other movements with positive or ambiguous signs. In the classical image, error does not express what is by right the worst that can happen to thought, without thought being presented as "willing" truth, as orientated toward truth, as turned toward truth. It is this confidence, which is not without humor, which animates the classical image—a relationship to truth that constitutes the infinite movement of knowledge as diagrammatic feature. In contrast, in the eighteenth century, what manifests the mutation of light from "natural light" to the "Enlightened" is the substitution of *belief* for knowledge—that is, a new infinite movement implying another image of thought: it is no longer a matter of turning toward but rather one of following tracks, of inferring rather than grasping or being grasped. Under what conditions is inference legitimate? Under what conditions can belief be legitimate when it has become secular? This question will be answered only with the creation of the great empiricist concepts (association, relation, habit, probability, convention). But conversely, these concepts, including the concept of belief itself, presuppose diagrammatic features that make belief an infinite movement independent of religion and traversing the new plane of immanence (religious belief, on the other hand, will become a conceptualizable case, the legitimacy or illegitimacy of which can be measured in accordance with the order of the infinite). Of course, we find in Kant many of these features inherited from Hume, but again at the price of a profound mutation, on a new plane or according to another image. Each time there are great acts of daring. When the distribution of what is due to thought by right changes, what changes from one plane of immanence to another are not only the positive or negative features but also the ambiguous features that may become increasingly numerous and that are no longer restricted to folding in accordance with a vectorial opposition of movements.

If we attempt to set out the features of a modern image of thought in such a summary fashion, this is not in a triumphalist way, or even in horror. No image of thought can be limited to a selection of calm determinations, and all of them encounter something that is abominable in principle, whether this be the error into which thought continually falls, or the illusion within which it continually turns, or the stupidity in which it continually wallows, or the delirium in which it continually turns away from itself or from a god. The Greek image of thought already invoked the madness of the double turning-away, which launched thought into infinite wandering rather than into error. The relationship of thought to truth in the ambiguities of infinite movement has never been a simple, let alone constant, matter. That is why it is pointless to rely on such a relationship to define philosophy. The first characteristic of the modern image of thought is, perhaps, the complete renunciation of this relationship so as to regard truth as solely the creation of thought, taking into account the plane of immanence that it takes as its presupposition, and all this plane's features, negative as well as positive having become indiscernible. As Nietzsche succeeded in making us understand, thought is creation, not will to truth. But if, contrary to what seemed to be the case in the classical image, there is no will to truth, this is because thought constitutes a simple "possibility" of thinking without yet defining a thinker "capable" of it and able to say "I": what violence must be exerted on thought for us to become capable of thinking; what violence of an infinite movement that, at the same time, takes from us our power to say "I"? Famous texts of Heidegger and Blanchot deal with this second characteristic. But, as a third characteristic, if there is in this way an "Incapacity" of thought, which remains at its core even after it has acquired the capacity determinable as creation, then a set of ambiguous signs arise, which become diagrammatic features or infinite movements and which take on a value by right, whereas in the other images of thought they were simple, derisory facts excluded from selection: as Kleist or Artaud suggests, thought as such begins to exhibit snarls, squeals, stammers; it talks in tongues and screams, which leads it to create, or to try to.[13] If thought searches, it is less in the manner of someone who possesses a method than that of a dog that seems to be making uncoordinated leaps. We have no reason to take pride in this image of thought, which involves much suffering without glory and indicates the degree to which thinking has become increasingly difficult: immanence.

The history of philosophy is comparable to the art of the portrait. It is not a matter of "making lifelike," that is, of repeating what a philosopher said but rather of producing resemblance by separating out both the plane of immanence he instituted and the new concepts he created. These are mental, noetic, and machinic portraits. Although they are usually created with philosophical tools, they can also be produced aesthetically. Thus Tinguely recently presented some monumental machinic portraits of philosophers, working with powerful, linked or alternating, infinite movements that can be folded over or spread out, with sounds, lightning flashes, substances of being, and images of thought according to complex curved planes.[14] However, if it is permissible to criticize such a great artist, the attempt does not quite seem to hit the mark. Nothing dances in the Nietzsche, although elsewhere Tinguely has been quite able to make machines dance. The Schopenhauer gives us nothing decisive, whereas the four Roots and the veil of Maya seem ready to occupy the bifaceted plane of the World as will and representation. The Heidegger does not retain any veiling-unveiling on the plane of a thought that does not yet think. Perhaps more attention should be given to the plane of immanence laid out as abstract machine and to created concepts as parts of the machine. In this sense we could imagine a machinic portrait of Kant, illusions included (see schema).

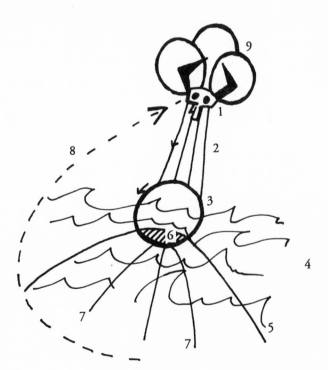

The components of the schema are as follows: 1) the "I think" as an ox head wired for sound, which constantly repeats Self = Self; 2) the categories as universal concepts (four great headings): shafts that are extensive and retractile according to the movement of 3); 3) the moving wheel of the schemata; 4) the shallow stream of Time as form of interiority, in and out of which the wheel of the schemata plunges; 5) space as form of exteriority: the stream's banks and bed; 6) the passive self at the bottom of the stream and as junction of the two forms; 7) the principles of synthetic judgments that run across space-time; 8) the transcendental field of possible experience, immanent to the "I" (plane of immanence); and 9) the three Ideas or illusions of transcendence (circles turning on the absolute horizon: Soul, World and God).

This account gives rise to many problems that concern philosophy and the history of philosophy equally. Sometimes the layers of the plane of immanence separate to the point of being opposed to one another, each one suiting this or that philosopher. Sometimes, on the contrary, they join together at least to cover fairly long periods. Moreover, the relationships between the instituting of a prephilosophical plane and the creation of philosophical concepts are themselves complex. Over a long period philosophers can create new concepts while remaining on the same plane and presupposing the same image as an earlier philosopher whom they invoke as their master: Plato and the neo-Platonists, Kant and the neo-Kantians (or even the way in which Kant himself reactivates certain parts of Platonism). However, in every case, this involves extending the original plane by giving it new curves, until a doubt arises: is this not a different plane that is woven in the mesh of the first one? Thus, the question of knowing when and to what extent philosophers are "disciples" of another philosopher and, on the contrary, when they are carrying out a critique of another philosopher by changing the plane and drawing up another image involves all the more complex and relative assessments, because the concepts that come to occupy a plane can never be simply deduced. Concepts that happen to populate a single plane, albeit at quite different times and with special connections, will be called concepts of the same group. Those concepts that refer back to different planes will not belong to the same group. There is a strict correspondence between the created concepts and the instituted plane, but this comes about through indirect relationships that are still to be determined.

Can we say that one plane is "better" than another or, at least, that it does or does not answer to the requirements of the age? What does answering to the requirements of the age mean, and what relationship is there between the movements or diagrammatic features of an image of thought and the movements or sociohistorical features of an age? We can only make headway with these questions if we give up the narrowly historical point of view of before and after in order to consider the time rather than the history of philosophy. This is a *stratigraphic* time where "before" and "after" indicate only an order of superimpositions. Certain paths (movements) take on sense and direction only as the shortcuts or detours of faded paths; a variable curvature can appear only as the transformation of one or more others; a stratum or layer of the plane of immanence will necessarily be *above* or *below* in relation to another, and images of thought cannot arise in any order whatever because they involve changes of orientation that can be directly located only on the earlier image (and even the point of condensation that determines the concept sometimes presupposes the breaking-up of a point or the conglomeration of earlier points). Mental landscapes do not change haphazardly through the ages: a mountain had to rise here or a river to flow by there again recently for the ground, now dry and flat, to have a particular appearance and texture. It is true that very old strata can rise to the surface again, can cut a path through the formations that covered them and surface directly on the current stratum to which they impart a new curvature. Furthermore, depending on the regions considered, superimpositions are not necessarily the same and do not have the same order. Philosophical time is thus a grandiose time of coexistence that does not exclude the before and after but *superimposes* them in a stratigraphic order. It is an infinite becoming of philosophy that crosscuts its history without being confused with it. The life of philosophers, and what is most external to their work, conforms to the ordinary laws of succession; but their proper names coexist and shine either as luminous points that take us through the components of a concept once more or as the cardinal points of a stratum or layer that continually come back to us, like dead stars whose light is brighter than ever. Philosophy is becoming, not history; it is the coexistence of planes, not the succession of systems.

That becoming, that coexistence is why planes may sometimes separate and sometimes join together—this is true for both the best and the worst. They have in common the restoration of transcendence and illusion (they cannot prevent it) but also the relentless struggle against transcendence and illusion; and each also has its particular way of doing both one and the other. Is there a "best" plane that would not hand over immanence to Something $= x$ and that would no longer mimic anything transcendent? We will say that *THE* plane of immanence is, at the same time, that which must be thought and that which cannot be thought. It is the nonthought within thought. It is the base of all planes, immanent to every thinkable plane that does not succeed in thinking it. It is the most intimate within thought and yet the absolute outside—an outside more distant than any external world because it is an inside deeper that any internal world: it is immanence, "intimacy as the Outside, the exterior become the intrusion that stifles, and the reversal of both the one and the other"[15]— the incessant to-ing and fro-ing of the plane, infinite movement. Perhaps this is the supreme act of philosophy: not so much to think *THE* plane of immanence as to show that it is there, unthought in every plane, and to think it in this way as the outside and inside of thought, as the not-external outside and the not-internal inside—that which cannot be thought and yet must be thought, which was thought once, as Christ was incarnated once, in order to show, that one time, the possibility of the impossible. Thus Spinoza is the Christ of philosophers, and the greatest philosophers are hardly more than apostles who distance themselves from or draw near to this mystery. Spinoza, the infinite becoming-philosopher: he showed, drew up, and thought the "best" plane of immanence—that is, the purest, the one that does not hand itself over to the transcendent or restore any transcendent, the one that inspires the fewest illusions, bad feelings, and erroneous perceptions.

3. Conceptual Personae

EXAMPLE 5

Although Descartes's cogito is created as a concept, it has presuppositions. This is not in the way that one concept presupposes others (for example, "man" presupposes "animal" and "rational"); the presuppositions here are implicit, subjective, and preconceptual, forming an image of thought: everyone knows what thinking means. Everyone can think; everyone wants the truth. Are these the only two elements—the concept and the plane of immanence or image of thought that will be occupied by concepts of the same group (the cogito and other concepts that can be connected to it)? Is there something else, in Descartes's case, other than the created cogito and the presupposed image of thought? Actually there is something else, somewhat mysterious, that appears from time to time or that shows through and seems to have a hazy existence halfway between concept and preconceptual plane, passing from one to the other. In the present case it is the Idiot: it is the Idiot who says "I" and sets up the cogito but who also has the subjective presuppositions or lays out the plane. The idiot is the private thinker, in contrast to the public teacher (the schoolman): the teacher refers constantly to taught concepts (man–rational animal), whereas the private thinker forms a concept with innate forces that everyone possesses on their own account by right ("I think"). Here is a very strange type of persona who wants to think, and who thinks for himself, by the "natural light." The idiot is a conceptual persona. The question "Are there precursors of the cogito?" can be made more precise. Where does the persona of the idiot come from, and how does it appear? Is it in a Christian atmosphere, but in reaction against the "scholastic" organization of Christianity and the authoritarian organization of the church? Can traces of this persona already be found in Saint Augustine? Is Nicholas of Cusa the one who accords the idiot full status as conceptual persona? This would be why he is close to the cogito but still unable to crystallize it as a concept.[1] In any case, the history of philosophy must go through these personae, through their changes according to planes and through their variety according to concepts. Philosophy constantly brings conceptual personae to life; it gives life to them.

The idiot will reappear in another age, in a different context that is still Christian, but Russian now. In becoming a Slav, the idiot is still the singular individual or private thinker, but with a different singularity. It is Chestov who finds in Dostoyevski the power of a new opposition between private thinker and public teacher.[2] The old idiot wanted indubitable truths at which he could arrive by himself: in the meantime he would doubt everything, even that $3 + 2 = 5$; he would doubt every truth of Nature. The new idiot has no wish for indubitable truths; he will never be "resigned" to the fact that $3 + 2 = 5$ and wills the absurd—this is not the same image of thought. The old idiot wanted truth, but the new idiot wants to turn the absurd into the highest power of thought—in other words, to create. The old idiot wanted to be accountable only to reason, but the new idiot, closer to Job than to Socrates, wants account to be taken of "every victim of History"—these are not the same concepts. The new idiot will never accept the truths of History. The old idiot wanted, by himself, to account for what was or was not comprehensible, what was or was not rational, what was lost or saved; but the new idiot wants the lost, the incomprehensible, and the absurd to be restored to him. This is most certainly not the same persona; a mutation has taken place. And yet a slender thread links the two idiots, as if the first had to lose reason so that the second rediscovers what the other, in winning it, had lost in advance: Descartes goes mad in Russia?

It is possible that the conceptual persona only rarely or allusively appears for himself. Nevertheless, he is there, and however nameless and subterranean, he must always be reconstituted by the reader. Sometimes he appears with a proper name: Socrates is the principal conceptual persona of Platonism. Many philosophers have written dialogues, but there is a danger of confusing the dialogue's characters with conceptual personae: they only nominally coincide and do not have the same role. The character of the dialogue sets out concepts: in the simplest case, one of the characters, who is sympathetic, is the author's representative; whereas the others, who are more-or-less antipathetic, refer to other philosophies whose concepts they expound in such a way as to prepare them for the criticisms or modifications to which the author wishes to subject them. On the other hand, conceptual personae carry out the movements that describe the author's plane of immanence, and they play a part in the very creation of the author's concepts. Thus, even when they are "antipathetic," they are so while belonging fully to the plane that the philosopher in question lays out and to the concepts that he creates. They then indicate the dangers specific to this plane, the bad perceptions, bad feelings, and even negative movements that emerge from it, and they will themselves inspire original concepts whose repulsive character remains a constitutive property of that philosophy. This is all the truer for the plane's *positive* movements, for *attractive* concepts and *sympathetic* personae: an entire philosophical *Einfühlung.** And in both cases there are often great ambiguities.

The conceptual persona is not the philosopher's representative but, rather, the reverse: the philosopher is only the envelope of his principal conceptual persona and of all the other personae who are the intercessors [*intercesseurs*], the real subjects of his philosophy. Conceptual personae are the philosopher's "heteronyms," and the philosopher's name is the simple pseudonym of his personae. I am no longer myself but thought's aptitude for finding itself and spreading across a plane that passes through me at several places. The philosopher is the idiosyncrasy of his conceptual personae. The destiny of the philosopher is to become his conceptual persona or personae, at the same time that these personae themselves become something other than what they are historically, mythologically, or commonly (the Socrates of Plato, the Dionysus of Nietzsche, the Idiot of Nicholas of Cusa). The conceptual persona is the becoming or the subject of a philosophy, on a par with the philosopher, so that Nicholas of Cusa, or even Descartes, should have signed themselves "the Idiot," just as Nietzsche signed himself "the Antichrist" or "Dionysus crucified." In everyday life speech-acts refer back to psychosocial types who actually attest to a subjacent third person: "I decree mobilization as President of the Republic," "I speak to you as father," and so on. In the same way, the philosophical shifter is a speech-act in the third person where it is always a conceptual persona who says "I": "I think as Idiot," "I will as Zarathustra," "I dance as Dionysus," "I claim as Lover." Even Bergsonian duration has need of a runner. In philosophical enunciations we do not do something by saying it but produce movement by thinking it, through the intermediary of a concep-

---

*Einfühlung, or empathy, as in the title of Wilhelm Worringer's great work *Abstraktion und Einfühlung,* translated into English as *Abstraction and Empathy* by Michael Bullock (London: Routledge, 1953).

tual persona. Conceptual personae are also the true agents of enunciation. "Who is 'I'?" It is always a third person.

We invoke Nietzsche because few philosophers have worked so much with both sympathetic (Dionysus, Zarathustra) and antipathetic (Christ, the Priest, the Higher Men; Socrates himself become antipathetic) conceptual personae. It might be thought that Nietzsche renounces concepts. However, he creates immense and intense concepts ("forces," "value," "becoming," "life"; and repulsive concepts like *ressentiment* and "bad conscience"), just as he lays out a new plane of immanence (infinite movements of the will to power and the eternal return) that completely changes the image of thought (criticism of the will to truth). But in Nietzsche, the conceptual personae involved never remain implicit. It is true that their manifestation for themselves gives rise to an ambiguity that leads many readers to see Nietzsche as a poet, thaumaturge, or creator of myths. But conceptual personae, in Nietzsche and elsewhere, are not mythical personifications or historical persons or literary or novelistic heroes. Nietzsche's Dionysus is no more the mythical Dionysus than Plato's Socrates is the historical Socrates. Becoming is not being, and Dionysus becomes philosopher at the same time that Nietzsche becomes Dionysus. Here, again, it is Plato who begins: he becomes Socrates at the same time that he makes Socrates become philosopher.

The difference between conceptual personae and aesthetic figures consists first of all in this: the former are the powers of concepts, and the latter are the powers of affects and percepts. The former take effect on a plane of immanence that is an image of Thought-Being (noumenon), and the latter take effect on a plane of composition as image of a Universe (phenomenon). The great aesthetic figures of thought and the novel but also of painting, sculpture, and music produce affects that surpass ordinary affections and perceptions, just as concepts go beyond everyday opinions. Melville said that a novel includes an infinite number of interesting characters but just one original Figure like the single sun of a constellation of a universe, like the beginning of things, or like the beam of light that draws a hidden universe out of the shadow: hence Captain Ahab, or Bartleby.[3] Kleist's universe is shot through with affects that traverse it like arrows or that suddenly freeze the universe in which the figures of Homburg or Penthesilea loom. Figures have nothing to do with resemblance or rhetoric but are the condition under which the arts produce affects of stone and metal, of strings and wind, of line and color, on a plane of composition of a universe. Art and philosophy crosscut the chaos and confront it, but it is not the same sectional plane; it is not populated in the same way. In the one there is the constellation of a universe or affects and percepts; and in the other, constitutions of immanence or concepts. Art thinks no less than philosophy, but it thinks through affects and percepts.

This does not mean that the two entities do not often pass into each other in a becoming that sweeps them both up in an intensity which co-determines them. With Kierkegaard, the theatrical and musical figure of Don Juan becomes a conceptual persona, and the Zarathustra persona is already a great musical and theatrical figure. It is as if, between them, not only alliances but also branchings and substitutions take place. In contemporary thought, Michel Guérin is one of those who has made the most profound discovery of the existence of conceptual personae at the heart of philosophy. But he defines them within a "logodrama" or a "figurology" that puts affect into thought.[4] This means that the concept as such can be concept of the affect, just as the affect can be affect of the concept. The plane of composition of art and the plane of immanence of philosophy can slip into each other to the degree that parts of one may be occupied by entities of the other. In fact, in each case the plane and that which occupies it are like two relatively distinct and heterogeneous parts. A thinker may therefore decisively modify what thinking means, draw up a new image of thought, and institute a new plane of immanence. But, instead of creating new concepts that occupy it, they populate it with other instances, with other poetic, novelistic, or even pictorial or musical entities. The opposite is also true. *Igitur* is just such a case of a conceptual persona transported onto a plane of composition, an aesthetic figure carried onto a plane of immanence: his proper name is a conjunction. These thinkers are "half" philosophers but also much more than philosophers. But they are not sages. There is such force in those unhinged works of Hölderlin, Kleist, Rimbaud, Mallarmé, Kafka, Michaux, Pessoa, Artaud, and many English and American novelists, from Melville to Lawrence or Miller, in which the reader discovers admiringly that they have written the novel of Spinozism. To be sure, they do not produce a synthesis of art and philosophy. They branch out and do not stop branching out. They are hybrid geniuses who neither erase nor cover over differences in kind but, on the contrary, use all the resources of their "athleticism" to install themselves within this very difference, like acrobats torn apart in a perpetual show of strength.

There is all the more reason for saying that conceptual personae (and also aesthetic figures) are irreducible to *psychosocial types,* even if here again there are constant penetrations. Simmel, and then Goffman, have probed far into the enclaves or margins of a society the study of these types, which often seem to be unstable: the stranger, the exile, the migrant, the transient, the native, the homecomer.[5] This is not through a taste for the anecdote. It seems to us that a social field comprises structures and functions, but this does not tell us very much directly about particular movements that affect the Socius. We already know the importance in animals of those activities that consist in forming *territories,* in abandoning or leaving them, and even in re-creating territory on something of a different nature (ethologists say that an animal's partner or friend is the "equivalent of a home" or that the family is a "mobile territory"). All the more so for the hominid: from its act of birth, it deterritorializes its front paw, wrests it from the earth to turn it into a hand, and reterritorializes it on branches and tools. A stick is, in turn, a deterritorialized branch. We need to see how everyone, at every age, in the smallest things as in the greatest challenges, seeks a territory, tolerates or carries out deterritorializations, and is reterritorialized on almost anything— memory, fetish, or dream. Refrains express these powerful dynamisms: my cabin in Canada . . . farewell, I am leaving . . . yes, it's me; I had to come back. We cannot even say what comes first, and perhaps every territory presupposes a prior deterritorialization, or everything happens at the same time. Social fields are inextricable knots in which the three movements are mixed up so that, in order to disentangle them, we have to *diagnose real types or personae*. The merchant buys in a territory, deterritorializes products into commodities, and is reterritorialized on commercial circuits. In capitalism, capital or property is deterritorialized, ceases to be landed, and is reterritorialized on the means of production; whereas labor becomes "abstract" labor, reterritorialized in wages: this is why Marx not only speaks of capital and labor but feels the need to draw up some true psychosocial types, both antipathetic and sympathetic: *the* capitalist, *the* proletarian. If we are looking for the originality of the Greek world we must ask what sort of territory is instituted by the Greeks, how they deterritorialize themselves, on what they are reterritorialized—and, in order to do this, to pick out specifically Greek types (the Friend, for example?). It is not always easy to decide which, at a given moment in a given society, are the good types: thus, the freed slave as type of deterritorialization in the Chinese Chou empire, the figure of the Exiled, of which the sinologist Tokei has given us a detailed portrait. We believe that psychosocial types have this meaning: to make perceptible, in the most insignificant or most important circumstances, the formation of territories, the vectors of deterritorialization, and the process of reterritorialization.

But are there not also territories and deterritorializations that are not only physical and mental but spiritual—not only relative but absolute in a sense yet to be determined? What is the Fatherland or Homeland invoked by the thinker, by the philosopher or artist? Philosophy is inseparable from a Homeland to which the a priori, the innate, or the memory equally attest. But why is this fatherland unknown, lost, or forgotten, turning the thinker into an Exile? What will restore an equivalent of territory, valid as a home? What will be philosophical refrains? What is thought's relationship with the earth? Socrates the Athenian, who does not like to travel, is guided by Parmenides of Elea when he is young, who is replaced by the Stranger when he is old, as if Platonism needed at least two conceptual personae.[6] What sort of stranger is there within the philosopher, with his look of returning from the land of the dead? *The role of conceptual personae is to show thought's territories, its absolute deterritorializations and reterritorializations.* Conceptual personae are thinkers, solely thinkers, and their personalized features are closely linked to the diagrammatic features of thought and the intensive features of concepts. A particular conceptual persona, who perhaps did not exist before us, thinks in us. For example, if we say that a conceptual persona stammers, it is no longer a type who stammers in a particular language but a thinker who makes the whole of language stammer: the interesting question then is "What is this thought that can only stammer?" Or again, if we say that a conceptual persona is the Friend, or that he is the Judge or the Legislator, we are no longer concerned with private, public, or legal status but with that which belongs by right to thought and only to thought. Stammerer, friend, or judge do not lose their concrete existence but, on the contrary, take on a new one as thought's internal conditions for its real exercise with this or that conceptual persona. This is not two friends who engage in thought; rather, it is thought itself that requires the thinker to be a friend so that thought is divided up within itself and can be exercised. It is thought itself which requires this division of thought between friends. These are no longer empirical, psychological, and social determinations, still less abstractions, but intercessors, crystals, or seeds of thought.

Even if the word *absolute* turns out to be exact, we must not think that deterritorializations and reterritorializations of thought transcend psychosocial ones, any more than they are reducible to them, or to an abstraction or ideological expression of them. Rather, there is a conjunction, a system of referrals or perpetual relays. The features of conceptual personae have relationships with the epoch or historical milieu in which they appear that only psychosocial types enable us to assess. But, conversely, the physical and mental movements of psychosocial types, their pathological symptoms, their relational attitudes, their existential modes, and their legal status, become susceptible to a determination purely of thinking and of thought that wrests them from both the historical state of affairs of a society and the lived experience of individuals, in order to turn them into the features of conceptual personae, or *thought-events* on the plane laid out by thought or under the concepts it creates. Conceptual personae and psychosocial types refer to each other and combine without ever merging.

No list of the features of conceptual personae can be exhaustive, since they are constantly arising and vary with planes of immanence. On a given plane, different kinds of features are mixed together to make up a persona. We assume there are *pathic features:* the Idiot, the one who wants to think for himself and is a persona who can change and take on another meaning. But also a Madman, a kind of madman, a cataleptic thinker or "mummy" who discovers in thought an inability to think; or a great maniac, someone frenzied, who is in search of that which precedes thought, an Already-there, but at the very heart of thought itself. Philosophy and schizophrenia have often been associated with each other. But in one case the schizophrenic is a conceptual persona who lives intensely within the thinker and forces him to think, whereas in the other the schizophrenic is a psychosocial type who represses the living being and robs him of his thought. Sometimes the two are combined, clasped together as if an event that is too intense corresponds to a lived condition that is too hard to bear.

There are *relational features:* "the Friend," but a friend who has a relationship with his friend only through the thing loved, which brings rivalry. The "Claimant" and the "Rival" quarrel over the thing or concept, but the concept needs a dormant, unconscious perceptible body, the "Boy" who is added to the conceptual personae. Are we not already on another plane, for love is like the violence that compels thinking—"Socrates the lover"—whereas friendship asks only for a little goodwill? And how could a "Fiancée" be denied her place in the role of conceptual persona, although it may mean rushing to her destruction, but not without the philosopher himself "becoming" woman? As Kierkegaard asks (or Kleist, or Proust): is not a woman more worthwhile than the friend who knows one well? And what happens if the woman herself becomes philosopher? Or a "Couple" who would be internal to thought and make "Socrates the husband" the conceptual persona? Unless we are led back to the "Friend," but after an ordeal that is too powerful, an inexpressible catastrophe, and so in yet another new sense, in a mutual distress, a mutual weariness that forms a new right of thought (Socrates becomes Jewish). Not two friends who communicate and recall the past together but, on the contrary, who suffer an amnesia or aphasia capable of splitting thought, of dividing it in itself. Personae proliferate and branch off, jostle one another and replace each other.[7]

There are *dynamic features:* if moving forward, climbing, and descending are dynamisms of conceptual personae, then leaping like Kierkegaard, dancing like Nietzsche, and diving like Melville are others for philosophical athletes irreducible to one another. And if today our sports are completely changing, if the old energy-producing activities are giving way to exercises that, on the contrary, insert themselves on existing energetic networks, this is not just a change in the type but yet other dynamic features that enter a thought that "slides" with new substances of being, with wave or snow, and turn the thinker into a sort of surfer as conceptual persona: we renounce then the energetic value of the sporting type in order to pick out the pure dynamic difference expressed in a new conceptual persona.

There are *juridical features* insofar as thought constantly lays claim to what belongs to it by right and, from the time of the pre-Socratics, has confronted Justice. But is this the power of the Claimant, or even of the Plaintiff, as philosophy extracts it from the tragic Greek tribunal? And will not philosophers be banned for a long time from being Judges, being at most doctors enrolled in God's justice, so long as they are not themselves the accused? When Leibniz turns the philosopher into the Lawyer of a god who is threatened on all sides, is this a new conceptual persona? Or the strange persona of Investigator advanced by the empiricists? It is Kant who finally turns the philosopher into the Judge at the same time that reason becomes a tribunal; but is this the legislative power of a determining judge, or the judicial power, the jurisprudence, of a reflecting judge? These are two quite different conceptual personae. Or else thought reverses everything—judges, lawyers, plaintiffs, accusers, and accused—like Alice on a plane of immanence where Justice equals Innocence, and where the Innocent becomes the conceptual persona who no longer has to justify herself, a sort of child-player against whom we can no longer do anything, a Spinoza who leaves no illusion of transcendence remaining. Should not judge and innocent merge into each other, that is to say, should not beings be judged from within—not at all in the name of the Law or of Values or even by virtue of their conscience but by the purely immanent criteria of their existence ("at all events, beyond Good and Evil does not mean beyond good and bad")?

And there are *existential features:* Nietzsche said that philosophy invents modes of existence or possibilities of life. That is why a few vital anecdotes are sufficient to produce a portrait of a philosophy, like the one Diogenes Laertius knew how to produce by writing the philosophers' bedside book or golden legend—Empedocles and his volcano, Diogenes and his barrel. It will be argued that most philosophers' lives are very bourgeois: but is not Kant's stocking-suspender a vital anecdote appropriate to the system of Reason?[8] And Spinoza's liking for battles between spiders is due to the fact that in a pure fashion they reproduce relationships of modes in the system of the *Ethics* as higher ethology. These anecdotes do not refer simply to social or even psychological types of philosopher (Empedocles the prince, Diogenes the slave) but show rather the conceptual personae who inhabit them. Possibilities of life or modes of existence can be invented only on a plane of immanence that develops the power of conceptual personae. The face and body of philosophers shelter these personae who often give them a strange appearance, especially in the glance, as if someone else was looking through their eyes. Vital anecdotes recount a conceptual persona's relationship with animals, plants, or rocks, a relationship according to which philosophers themselves become something unexpected and take on a tragic and comic dimension that they could not have by themselves. It is through our personae that we philosophers become always something else and are reborn as public garden or zoo.

EXAMPLE 6

Even illusions of transcendence are useful to us and provide vital anecdotes—for when we take pride in encountering the transcendent within immanence, all we do is recharge the plane of immanence with immanence itself: Kierkegaard leaps outside the plane, but what is "restored" to him in this suspension, this halted movement, is the fiancée or the lost son, it is existence on the plane of immanence.[9] Kierkegaard does not hesitate to say so: a little "resignation" will be enough for what belongs to transcendence, but *immanence must also be restored*. Pascal wagers on the transcendent existence of God, but the stake, that *on* which one bets, is the immanent existence of the one who believes that God exists. Only that existence is able to cover the plane of immanence, to achieve infinite movement, and to produce and reproduce intensities; whereas the existence of the one who does not believe that God exists falls into the negative. It might even be said here, as François Jullien says of Chinese thought, that transcendence is relative and represents no more than an "absolutization of immanence."[10] There is not the slightest reason for thinking that modes of existence need transcendent values by which they could be compared, selected, and judged relative to one another. On the contrary, there are only immanent criteria. A possibility of life is evaluated through itself in the movements it lays out and the intensities it creates on a plane of immanence: what is not laid out or created is rejected. A mode of existence is good or bad, noble or vulgar, complete or empty, independently of Good and Evil or any transcendent value: there are never any criteria other than the tenor of existence, the intensification of life. Pascal and Kierkegaard, who were familiar with infinite movements, and who extracted from the Old Testament new conceptual personae able to stand up to Socrates, were well aware of this. Kierkegaard's "knight of the faith," he who makes the leap, or Pascal's gambler, he who throws the dice, are men of a transcendence or a faith. But they constantly recharge immanence: they are philosophers or, rather, intercessors, conceptual personae who stand in for these two philosophers and who are concerned no longer with the transcendent existence of God but only with the infinite immanent possibilities brought by the one who believes that God exists.

The problem would change if it were another plane of immanence. It is not that the person who does not believe God exists would gain the upper hand, since he would still belong to the old plane as negative movement. But, on the new plane, it is possible that the problem now concerns the one who believes in the world, and not even in the existence of the world but in its possibilities of movements and intensities, so as once again to give birth to new modes of existence, closer to animals and rocks. It may be that believing in this world, in this life, becomes our most difficult task, or the task of a mode of existence still to be discovered on our plane of immanence today. This is the empiricist conversion (we have so many reasons not to believe in the human world; we have lost the world, worse than a fiancée or a god). The problem has indeed changed.

The conceptual persona and the plane of immanence presuppose each other. Sometimes the persona seems to precede the plane, sometimes to come after it—that is, it appears twice; it intervenes twice. On the one hand, it plunges into the chaos from which it extracts the determinations with which it produces the diagrammatic features of a plane of immanence: it is as if it seizes a handful of dice from chance-chaos so as to throw them on a table. On the other hand, the persona establishes a correspondence between each throw of the dice and the intensive features of a concept that will occupy this or that region of the table, as if the table were split according to the combinations. Thus, the conceptual persona with its personalized features intervenes between chaos and the diagrammatic features of the plane of immanence and also between the plane and the intensive features of the concepts that happen to populate it: *Igitur*. Conceptual personae constitute points of view according to which planes of immanence are distinguished from one another or brought together, but they also constitute the conditions under which each plane finds itself filled with concepts of the same group. Every thought is a Fiat, expressing a throw of the dice: constructivism. But this is a very complex game, because throwing involves infinite movements that are reversible and folded within each other so that the consequences can only be produced at infinite speed by creating finite forms corresponding to the intensive ordinates of these movements: every concept is a combination that did not exist before. Concepts are not deduced from the plane. The conceptual persona is needed to create concepts on the plane, just as the plane itself needs to be laid out. But these two operations do not merge in the persona, which itself appears as a distinct operator.

There are innumerable planes, each with a variable curve, and they group together or separate themselves according to the points of view constituted by personae. Each persona has several features that may give rise to other personae, on the same or a different plane: conceptual personae proliferate. There is an infinity of possible concepts on a plane: they resonate and connect up with mobile bridges, but it is impossible to foresee the appearance they take on as a function of variations of curvature. They are created in bursts and constantly bifurcate. The game is all the more complex because on each plane *negative* movements are enveloped within positive movements, expressing the risks and dangers confronted by thought, the false perceptions and bad feelings that surround it. There are also *antipathetic* conceptual personae who cling to sympathetic personae and from whom the latter do not manage to free themselves (it is not only Zarathustra who is haunted by "his" ape or clown, or Dionysus who does not separate himself from Christ; but Socrates who never manages to distinguish himself from "his" sophist, and the critical philosopher who is always warding off his bad doubles). Finally, there are *repulsive* concepts locked within attractive ones but that outline regions of low or empty intensity on the plane and that continually cut themselves off, create discordancies, and sever connections (does not transcendence itself have "its" concepts?). But even more than a vectorial distribution, the signs, personae, and concepts of planes are ambiguous because they are folded within one another, embrace or lie alongside one another. That is why philosophy always works blow by blow.

Philosophy presents three elements, each of which fits with the other two but must be considered for itself: *the prephilosophical plane it must lay out (immanence), the persona or personae it must invent and bring to life (insistence), and the philosophical concepts it must create (consistency)*. Laying out, inventing, and creating constitute the philosophical trinity—diagrammatic, personalistic, and intensive features. Concepts are grouped according to whether they resonate or throw out mobile bridges, covering the same plane of immanence that connects them to one another. There are families of planes according to whether the infinite movements of thought fold within one another and compose variations of curvature or, on the contrary, select noncomposable varieties. There are types of persona according to the possibilities of even their hostile encounters on the same plane and in a group. But it is often difficult to determine if it is the same group, the same type, or the same family. A whole "taste" is needed here.

Since none of these elements are deduced from the others, there must be coadaptation of the three. The philosophical faculty of coadaptation, which also regulates the creation of concepts, is called *taste*. If the laying-out of the plane is called Reason, the invention of personae Imagination, and the creation of concepts Understanding, then taste appears as the triple faculty of the still-undetermined concept, of the persona still in limbo, and of the still-transparent plane. That is why it is necessary to create, invent, and lay out, while taste is like the rule of correspondence of the three instances that are different in kind. It is certainly not a faculty of measuring. No measure will be found in those infinite movements that make up the plane of immanence, in those accelerated lines without contour, and those inclines and curves; or in those always excessive and sometimes antipathetic personae; or in those concepts with irregular forms, strident intensities, and colors that are so bright and barbarous that they can inspire a kind of "disgust" (especially in repulsive concepts). Nevertheless, what appears as philosophical taste in every case is love of the well-made concept, "well-made" meaning not a moderation of the concept but a sort of stimulation, a sort of modulation in which conceptual activity has no limit in itself but only in the other two limitless activities. If ready-made concepts already existed they would have to abide by limits. But even the "prephilosophical" plane is only so called because it is laid out as presupposed and not because it preexists without being laid out. The three activities are strictly simultaneous and have only incommensurable relationships. The creation of concepts has no other limit than the plane they happen to populate; but the plane itself is limitless, and its layout only conforms to the concepts to be created that it must connect up, or to the personae to be invented that it must maintain. It is as in painting: there is a taste according to which even monsters and dwarves must be well made, which does not mean insipid but that their irregular contours are in keeping with a skin texture or with a background of the earth as germinal substance with which they seem to fit. There is a taste for colors that, in great painters, does not result in restraint in the creation of colors but, on the contrary, drives them to the point where colors encounter their figures made of contours, and their plane made of flats, curves, and arabesques. Van Gogh takes yellow to the limitless only by inventing the man-sunflower and by laying out the plane of infinite little commas. The taste for colors shows at once the respect with which they must be approached, the long wait that must be passed through, but also the limitless creation that makes them exist. The same goes for the taste for concepts: the philosopher does not approach the undetermined concept except with fear and respect, and he hesitates for a long time before setting forth; but he can determine a concept only through a measureless creation whose only rule is a plane of immanence that he lays out and whose only compass are the strange personae to which it gives life. Philosophical taste neither replaces creation nor restrains it. On the contrary, the creation of concepts calls for a taste that modulates it. The free creation of determined concepts needs a taste for the undetermined concept. Taste is this power, this being-potential of the concept: it is certainly not for "rational or reasonable" reasons that a particular concept is created or a particular component chosen. Nietzsche sensed this relationship of the creation of concepts with a specifically philosophical taste, and if the philosopher is he who creates concepts, it is thanks to a faculty of taste that is like an instinctive, almost animal *sapere*—a *Fiat* or a *Fatum* that gives each philosopher the right of access to certain problems, like an imprint on his name or an affinity from which his works flow.[11]

A concept lacks meaning to the extent that it is not connected to other concepts and is not linked to a problem that it resolves or helps to resolve. But it is important to distinguish philosophical from scientific problems. Little is gained by saying that philosophy asks "questions," because *question* is merely a word for problems that are irreducible to those of science. Since concepts are not propositional, they cannot refer to problems concerning the extensional conditions of propositions assimilable to those of science. If, all the same, we continue to translate the philosophical concept into propositions, this can only be in the form of more-or-less plausible opinions without scientific value. But in this way we encounter a difficulty that the Greeks had already come up against. This is the third characteristic by which philosophy is thought of as something Greek: the Greek city puts forward the friend or rival as social relation, and it lays out a plane of immanence—but it also makes *free opinion* (*doxa*) prevail. Philosophy must therefore extract from opinions a "knowledge" that transforms them but that is also distinct from science. The philosophical problem thus consists in finding, in each case, the instance that is able to gauge a truth value of opposable opinions, either by selecting some as more wise than others or by fixing their respective share of the truth. Such was always the meaning of what is called dialectic and that reduces philosophy to interminable discussion.[12] This can be seen in Plato, where universals of contemplation are supposed to gauge the respective value of rival opinions so as to raise them to the level of knowledge. It is true that there are still contradictions in Plato, in the so-called aporetic dialogues, which forced Aristotle to direct the dialectical investigation of problems toward universals of communication (the topics). In Kant, again, the problem will consist in the selection or distribution of opposed opinions, but thanks to

universals of reflection, until Hegel has the idea of making use of the
contradiction between rival opinions to extract from them suprascien-
tific propositions able to move, contemplate, reflect, and communicate in themselves and within the absolute (the speculative proposition wherein opinions become moments of the concept). But, beneath
the highest ambitions of the dialectic, and irrespective of the genius
of the great dialecticians, we fall back into the most abject conditions
that Nietzsche diagnosed as the art of the pleb or bad taste in philosophy: a reduction of the concept to propositions like simple opinions;
false perceptions and bad feelings (illusions of transcendence or of
universals) engulfing the plane of immanence; the model of a form of
knowledge that constitutes only a supposedly higher opinion, *Urdoxa;* a replacement of conceptual personae by teachers or leaders of
schools. The dialectic claims to discover a specifically philosophical
discursiveness, but it can only do this by linking opinions together.
It has indeed gone beyond opinion toward knowledge, but opinion
breaks through and continues to break through. Even with the resources of an *Urdoxa,* philosophy remains a doxography. It is always
the same melancholy that raises disputed Questions and Quodlibets
from the Middle Ages where one learns what each doctor thought
without knowing why he thought it (the Event), and that one finds
again in many histories of philosophy in which solutions are reviewed
without ever determining what the problem is (substance in Aristotle, Descartes, Leibniz), since the problem is only copied from the
propositions that serve as its answer.

If philosophy is paradoxical by nature, this is not because it sides
with the least plausible opinion or because it maintains contradictory
opinions but rather because it uses sentences of a standard language
to express something that does not belong to the order of opinion or
even of the proposition. The concept is indeed a solution, but the
problem to which it corresponds lies in its intensional conditions of
consistency and not, as in science, in the conditions of reference of
extensional propositions. If the concept is a solution, the conditions

of the philosophical problem are found on the plane of immanence presupposed by the concept (to what infinite movement does it refer in the image of thought?), and the unknowns of the problem are found in the conceptual personae that it calls up (what persona, exactly?). A concept like knowledge has meaning only in relation to an image of thought to which it refers and to a conceptual persona that it needs; a different image and a different persona call for other concepts (belief, for example, and the Investigator). A solution has no meaning independently of a problem to be determined in its conditions and unknowns; but these conditions and unknowns have no meaning independently of solutions determinable as concepts. Each of the three instances is found in the others, but they are not of the same kind, and they coexist and subsist without one disappearing into the other. Bergson, who contributed so much to the comprehension of the nature of philosophical problems, said that a well-posed problem was a problem solved. But this does not mean that a problem is merely the shadow or epiphenomenon of its solutions, or that the solution is only the redundancy or analytical consequence of the problem. Rather, the three activities making up constructionism continually pass from one to the other, support one another, sometimes precede and sometimes follow each other, one creating concepts as a case of solution, another laying out a plane and a movement on the plane as the conditions of a problem, and the other inventing a persona as the unknown of the problem. The whole of the problem (of which the solution is itself a part) always consists in constructing the other two when the third is underway. We have seen how, from Plato to Kant, thought, "first," and time took different concepts that were able to determine solutions, but on the basis of presuppositions that determined different problems. This is because the same terms can appear twice and even three times: once in solutions as concepts, again in the presupposed problems, and once more in a persona as intermediary, intercessor. But each time it appears in a specific, irreducible form.

No rule, and above all no discussion, will say in advance whether this is the good plane, the good persona, or the good concept; for each of them determines if the other two have succeeded or not, but each must be constructed on its own account—one created, one invented, and the other laid out. Problems and solutions are constructed about which we can say, "Failure ... Success ... ," but only as we go along and on the basis of their coadaptations. Constructivism disqualifies all discussion—which holds back the necessary constructions—just as it exposes all the universals of contemplation, reflection, and communication as sources of what are called "false problems" emanating from the illusions surrounding the plane. That is all that can be said in advance. It is possible that we think we have found a solution; but a new curve of the plane, which at first we did not see, starts it all off again, posing new problems, a new batch of problems, advancing by successive surges and seeking concepts to come, concepts yet to be created (we do not even know if this is not a new plane that has separated from the preceding plane). Conversely, it is possible that a new concept is buried like a wedge between what one thought were two neighboring concepts, seeking in its turn the determination of a problem that appears like a sort of extension on the table of immanence. Philosophy thus lives in a permanent crisis. The plane takes effect through shocks, concepts proceed in bursts, and personae by spasms. The relationship among the three instances is problematic by nature.

We cannot say in advance whether a problem is well posed, whether a solution fits, is really the case, or whether a persona is viable. This is because the criteria for each philosophical activity are found only in the other two, which is why philosophy develops in paradox. Philosophy does not consist in knowing and is not inspired by truth. Rather, it is categories like Interesting, Remarkable, or Important that determine success or failure. Now, this cannot be known before being constructed. We will not say of many books of philosophy that they are false, for that is to say nothing, but rather that they lack importance or interest, precisely because they do not create any concept or contribute an image of thought or beget a persona worth the effort. Only teachers can write "false" in the margins, perhaps; but readers doubt the importance and interest, that is to say, the novelty of what they are given to read. These are categories of the Mind. Melville said that great novelistic characters must be Originals, Unique. The same is true of conceptual personae. They must be remarkable, even if they are antipathetic; a concept must be interesting, even if it is repulsive. When Nietzsche constructed the concept of "bad conscience" he could see in this what is most disgusting in the world and yet exclaim, "This is where man begins to be interesting!" and consider himself actually to have created a new concept for man, one that suited man, related to a new conceptual persona (the priest) and with a new image of thought (the will to power understood from the point of view of nihilism).[13]

Criticism implies new concepts (of the thing criticized) just as much as the most positive creation. Concepts must have irregular contours molded on their living material. What is naturally uninteresting? Flimsy concepts, what Nietzsche called the "formless and fluid daubs of concepts"—or, on the contrary, concepts that are too regular, petrified, and reduced to a framework. In this respect, the most universal concepts, those presented as eternal forms or values, are the most skeletal and least interesting. Nothing positive is done, nothing at all, in the domains of either criticism or history, when we are content to brandish ready-made old concepts like skeletons intended to intimidate any creation, without seeing that the ancient philosophers from whom we borrow them were already doing what we would like to prevent modern philosophers from doing: they were creating their concepts, and they were not happy just to clean and scrape bones like the critic and historian of our time. Even the history of philosophy is completely without interest if it does not undertake to awaken a dormant concept and to play it again on a new stage, even if this comes at the price of turning it against itself.

4. Geophilosophy

Subject and object give a poor approximation of thought. Thinking is neither a line drawn between subject and object nor a revolving of one around the other. Rather, thinking takes place in the relationship of territory and the earth. Kant is less a prisoner of the categories of subject and object than he is believed to be, since his idea of Copernican revolution puts thought into a direct relationship with the earth. Husserl demands a ground for thought as original intuition, which is like the earth inasmuch as it neither moves nor is at rest. Yet we have seen that the earth constantly carries out a movement of deterritorialization on the spot, by which it goes beyond any territory: it is deterritorializing and deterritorialized. It merges with the movement of those who leave their territory en masse, with crayfish that set off walking in file at the bottom of the water, with pilgrims or knights who ride a celestial line of flight. The earth is not one element among others but rather brings together all the elements within a single embrace while using one or another of them to deterritorialize territory. Movements of deterritorialization are inseparable from territories that open onto an elsewhere; and the process of reterritorialization is inseparable from the earth, which restores territories. Territory and earth are two components with two zones of indiscernibility—deterritorialization (from territory to the earth) and reterritorialization (from earth to territory). We cannot say which comes first. In what sense, we ask, is Greece the philosopher's territory or philosophy's earth?

States and Cities have often been defined as territorial, as substituting a territorial principle for the principle of lineage. But this is inexact: lineal groups may change territory, and they are only really determined by embracing a territory or residence in a "local lineage." State and City, on the contrary, carry out a deterritorialization because the former juxtaposes and compares agricultural territories by relating them to a higher arithmetical Unity, and the latter adapts the territory to a geometrical extensiveness that can be continued in commercial circuits. The *imperial spatium* of the State and the *political extensio* of the city are not so much forms of a territorial principle as a deterritorialization that takes place on the spot when the State appropriates the territory of local groups or when the city turns its back on its hinterland. In one case, there is reterritorialization on the palace and its supplies; and in the other, on the agora and commercial networks.

In imperial states deterritorialization takes place through transcendence: it tends to develop vertically from on high, according to a celestial component of the earth. The territory has become desert earth, but a celestial Stranger arrives to reestablish the territory or reterritorialize the earth. In the city, by contrast, deterritorialization takes place through immanence: it frees an Autochthon, that is to say, a power of the earth that follows a maritime component that goes under the sea to reestablish the territory (the Erechtheum, temple of Athena and Poseidon). In fact, things are more complicated because the imperial Stranger himself needs surviving Autochthons and because the citizen Autochthon calls on strangers in flight—but these are not at all the same psychosocial types, any more than the polytheism of the empire and the polytheism of the city are the same religious figures.[1]

Greece seems to have a fractal structure insofar as each point of the peninsula is close to the sea and its sides have great length. The Aegean peoples, the cities of ancient Greece and especially Autochthonous Athens, were not the first commercial cities. But they are the first to be at once near enough to and far enough away from the archaic eastern empires to be able to benefit from them without following their model. Rather than establish themselves in the pores of the empires, they are steeped in a new component; they develop a particular mode of deterritorialization that proceeds by immanence; they form a *milieu of immanence*. It is like an "international market" organized along the borders of the Orient between a multiplicity of independent cities or distinct societies that are nevertheless attached to one another and within which artisans and merchants find a freedom and mobility denied to them by the empires.[2] These types come from the borderlands of the Greek world, strangers in flight, breaking with empire and colonized by peoples of Apollo—not only artisans and merchants but philosophers. As Faye says, it took a century for the name *philosopher,* no doubt invented by Heraclitus of Ephesus, to find its correlate in the word *philosophy,* no doubt invented by Plato the Athenian: "Asia, Italy, and Africa are the odyssean phases of the journey connecting *philosophos* to philosophy."[3] Philosophers are strangers, but philosophy is Greek. What do these emigres find in the Greek milieu? At least three things are found that are the de facto conditions of philosophy: a pure sociability as milieu of immanence, the "intrinsic nature of association," which is opposed to imperial sovereignty and implies no prior interest because, on the contrary, competing interests presuppose it; a certain pleasure in forming associations, which constitutes friendship, but also a pleasure in breaking up the association, which constitutes rivalry (were there not already "societies of friends" formed by emigres, like the Pythagoreans, but still somewhat secret, which found their chance in Greece?); and a taste for opinion inconceivable in an empire, a taste for the exchange of views, for conversation.[4] We constantly rediscover these three Greek features: immanence, friendship, and opinion. We do not see a softer world here because sociability has its cruelties, friendship has its rivalries, and opinion has its antagonisms and bloody reversals. Salamis is the Greek miracle where Greece escapes from the Persian empire and where the autochthonous people who lost its territory prevails on the sea, is reterritorialized on the sea. The Delian League is like the fractalization of Greece. For a fairly short period the deepest bond existed between the democratic city, colonization, and a new imperialism that no longer saw the sea as a limit of its territory or an obstacle to its endeavor but as a wider bath of immanence. All of this, and primarily philosophy's link with Greece, seems a recognized fact, but it is marked by detours and contingency.

Whether physical, psychological, or social, deterritorialization is *relative* insofar as it concerns the historical relationship of the earth with the territories that take shape and pass away on it, its geological relationship with eras and catastrophes, its astronomical relationship with the cosmos and the stellar system of which it is a part. But deterritorialization is *absolute* when the earth passes into the pure plane of immanence of a Being-thought, of a Nature-thought of infinite diagrammatic movements. Thinking consists in stretching out a plane of immanence that absorbs the earth (or rather, "adsorbs" it). Deterritorialization of such a plane does not preclude reterritorialization but posits it as the creation of a future new earth. Nonetheless, absolute deterritorialization can only be thought according to certain still-to-be-determined relationships with relative deterritorializations that are not only cosmic but geographical, historical, and psychosocial. There is always a way in which absolute deterritorialization takes over from a relative deterritorialization in a given field.

It is at this point that a major difference arises depending on whether relative deterritorialization takes place through immanence or through transcendence. When it is transcendent, vertical, celestial, and brought about by the imperial unity, the transcendent element must always give way or submit to a sort of rotation in order to be inscribed on the always-immanent plane of Nature-thought. The celestial vertical settles on the horizontal of the plane of thought in accordance with a spiral. Thinking here implies a projection of the transcendent on the plane of immanence. Transcendence may be entirely "empty" in itself, yet it becomes full to the extent that it descends and crosses different hierarchized levels that are projected together on a region of the plane, that is to say, on an aspect corresponding to an infinite movement. In this respect, it is the same when transcendence invades the absolute or monotheism replaces unity: the transcendent God would remain empty, or at least *absconditus,* if it were not projected on a plane of immanence of creation where it traces the stages of its theophany. In both cases, imperial unity or spiritual empire, the transcendence that is projected on the plane of immanence paves it or populates it with Figures. It is a wisdom or a religion—it does not much matter which. It is only from this point of view that Chinese hexagrams, Hindu mandalas, Jewish sephiroth, Islamic "imaginals," and Christian icons can be considered together: thinking through figures. Hexagrams are combinations of continuous and discontinuous features deriving from one another according to the levels of a spiral that figures the set of moments through which the transcendent descends. The mandala is a projection on a surface that establishes correspondence between divine, cosmic, political, architectural, and organic levels as so many values of one and the same transcendence. That is why the figure has a reference, one that is plurivocal and circular by nature. Certainly, it is not defined by an external resemblance, which remains prohibited, but by an internal tension that relates it to the transcendent on the plane of immanence of thought. In short, the figure is essentially *paradigmatic, projective, hierarchical,* and *referential* (the arts and sciences also set up powerful figures, but what distinguishes them from all religion is not that they lay claim to prohibited resemblance but that they emancipate a particular level so as to make it into new planes of thought on which, as will be seen, the nature of the references and projections change).

Earlier, in order to move on quickly, we said that the Greeks invented an absolute plane of immanence. But the originality of the Greeks should rather be sought in the relation between the relative and the absolute. When relative deterritorialization is itself horizontal, or immanent, it combines with the absolute deterritorialization of the plane of immanence that carries the movements of relative deterritorialization to infinity, pushes them to the absolute, by transforming them (milieu, friend, opinion). Immanence is redoubled. This is where one thinks no longer with figures but with concepts. It is the concept that comes to populate the plane of immanence. There is no longer projection in a figure but connection in the concept. This is why the concept itself abandons all reference so as to retain only the conjugations and connections that constitute its consistency. The concept's only rule is internal or external neighborhood. Its internal neighborhood or consistency is secured by the connection of its components in zones of indiscernibility; its external neighborhood or exoconsistency is secured by the bridges thrown from one concept to another when the components of one of them are saturated. And this is really what the creation of concepts means: to connect internal, inseparable components to the point of closure or saturation so that we can no longer add or withdraw a component without changing the nature of the concept; to connect the concept with another in such a way that the nature of other connections will change. The plurivocity of the concept depends solely upon neighborhood (one concept can have several neighborhoods). Concepts are flat surfaces without levels, orderings without hierarchy; hence the importance in philosophy of the questions "What to put in a concept?" and "What to put with it?" What concept should be put alongside a former concept, and what components should be put in each? These are the questions of the creation of concepts. The pre-Socratics treat physical elements like concepts: they take them for themselves, independently of any reference, and seek only the good rules of neighborhood between them and in their possible components. If their answers vary it is because, inside and outside, they do not compose these elementary concepts in the same way. The concept is not paradigmatic but *syntagmatic;* not projective but *connective*; not hierarchical but *linking;** not referential but *consistent.* That being so, it is inevitable that philosophy, science, and art are no longer organized as levels of a single projection and are not even differentiated according to a common matrix but are immediately posited or reconstituted in a respective independence, in a division of labor that gives rise to relationships of connection between them.

Must we conclude from this that there is a radical opposition between figures and concepts? Most attempts to fix their differences express only ill-tempered judgments that are content to depreciate one or other of the terms: sometimes concepts are endowed with the prestige of reason while figures are referred to the night of the irrational and its symbols; sometimes figures are granted the privileges of spiritual life while concepts are relegated to the artificial movements of a dead understanding. And yet disturbing affinities appear on what seems to be a common plane of immanence.[5] In a sort of to-ing and fro-ing, Chinese thought inscribes the diagrammatic movements of a Nature-thought on the plane, yin and yang; and hexagrams are sections of the plane, intensive ordinates of these infinite movements, with their components in continuous and discontinuous features. But correspondences like these do not rule out there being a boundary, however difficult it is to make out. This is because figures are projections on the plane, which implies something vertical or transcendent. Concepts, on the other hand, imply only neighbor-

*I.e., *vicinal:* this term is usually used in French to describe a byroad or byway or a road that links together a number of villages and hamlets. "Linking" is not exact but conveys the appropriate contrast with *hierarchical.*

hoods and connections on the horizon. Certainly, as François Jullien has already shown in the case of Chinese thought, the transcendent produces an "absolutization of immanence" through projection. But philosophy appeals to a completely different immanence of the absolute. All that can be said is that figures tend toward concepts to the point of drawing infinitely near to them. From the fifteenth to the seventeenth century, Christianity made the *impresa* the envelope of a "concetto," but the concetto has not yet acquired consistency and depends upon the way in which it is figured or even dissimulated. The question that arises periodically—"Is there a Christian philosophy?"—means "Is Christianity able to create proper concepts?" (belief, anguish, sin, freedom). We have seen this in Pascal or Kierkegaard: perhaps belief becomes a genuine concept only when it is made into belief in this world and is connected rather than being projected. Perhaps Christianity does not produce concepts except through its atheism, through the atheism that it, more than any other religion, secretes. Atheism is not a problem for philosophers or the death of God. Problems begin only afterward, when the atheism of the concept has been attained. It is amazing that so many philosophers still take the death of God as tragic. Atheism is not a drama but the philosopher's serenity and philosophy's achievement. There is always an atheism to be extracted from a religion. This was already true in Jewish thought: it pushed its figures as far as the concept, but it arrived at that point only with the atheist Spinoza. And if it is true that figures tend toward concepts in this way, the converse is equally true, and philosophical concepts reproduce figures whenever immanence is attributed to something. The three figures of philosophy are objectality of contemplation, subject of reflection, and intersubjectivity of communication. It should be noted that religions do not arrive at the concept without denying themselves, just as philosophies do not arrive at the figure without betraying themselves. There is a difference of kind between figures and concepts, but every possible difference of degree also.

Can we speak of Chinese, Hindu, Jewish, or Islamic "philosophy"? Yes, to the extent that thinking takes place on a plane of immanence that can be populated by figures as much as by concepts. However, this plane of immanence is not exactly philosophical, but prephilosophical. It is affected by what populates and reacts on it, in such a way that it becomes philosophical only through the effect of the concept. Although the plane is presupposed by philosophy, it is nonetheless instituted by it and it unfolds in a philosophical relationship with the nonphilosophical. In the case of figures, on the other hand, the prephilosophical shows that a creation of concepts or a philosophical formation was not the inevitable destination of the plane of immanence itself but that it could unfold in wisdoms and religions according to a bifurcation that wards off philosophy in advance from the point of view of its very possibility. What we deny is that there is any internal necessity to philosophy, whether in itself or in the Greeks (and the idea of a Greek miracle would only be another aspect of this pseudonecessity). Nevertheless, philosophy was something Greek—although brought by immigrants. The birth of philosophy required an *encounter* between the Greek milieu and the plane of immanence of thought. It required the conjunction of two very different movements of deterritorialization, the relative and the absolute, the first already at work in immanence. Absolute deterritorialization of the plane of thought had to be aligned or directly connected with the relative deterritorialization of Greek society. The encounter between friend and thought was needed. In short, philosophy does have a principle, but it is a synthetic and contingent principle—an encounter, a conjunction. It is not insufficient by itself but contingent in itself. Even in the concept, the principle depends upon a connection of components that could have been different, with different neighborhoods. The principle of reason such as it appears in philosophy is a principle of contingent reason and is put like this: there is no good reason but contingent reason; there is no universal history except of contingency.

EXAMPLE 7

It is pointless to seek, like Hegel or Heidegger, an analytic and necessary principle that would link philosophy to Greece. Because the Greeks are free men they are the first to grasp the Object in a relationship with the subject: according to Hegel, this would be the concept. But, because the object is still *contemplated* as "beautiful," without its relationship to the subject yet being determined, we must await the following stages for this relationship to be *reflected* itself and then put into movement or *communicated*. Nonetheless it remains the case that the Greeks invented the first stage on the basis of which everything develops internally to the concept. No doubt the Orient thought, but it thought the object in itself as pure abstraction, the empty universality identical to simple particularity: it lacked the relationship to the subject as concrete universality or as universal individuality. The Orient is unaware of the concept because it is content to put the most abstract void and the most trivial being in a relationship of coexistence without any mediation. However, it is not clear what distinguishes the antephilosophical stage of the Orient and the philosophical stage of Greece, since Greek thought is not conscious of the relationship to the subject that it presupposes without yet being able to reflect.

Thus, Heidegger displaces the problem and situates the concept in the difference between Being and beings rather than in that between subject and object. He views the Greek as the Autochthon rather than as the free citizen (and, as the themes of building and dwelling indicate, all of Heidegger's reflection on Being and beings brings earth and territory together): the specificity of the Greek is to dwell in Being and to possess its word. Deterritorialized, the Greek is reterritorialized on his own language and its linguistic treasure—

the verb *to be*. Thus, the Orient is not before philosophy but alongside, because it thinks but it does not think Being.[6] Philosophy does not so much evolve and pass through degrees of subject and object as haunt a structure of Being. Heidegger's Greeks never succeed in "articulating" their relationship to Being; Hegel's Greeks never came to reflect their relationship to the Subject. But in Heidegger it is not a question of going farther than the Greeks; it is enough to resume their movement in an initiating, recommencing repetition. This is because Being, by virtue of its structure, continually turns away when it turns toward, and the history of Being or of the earth is the history of its turning away, of its deterritorialization in the technico-worldwide development of Western civilization started by the Greeks and reterritorialized on National Socialism. What remains common to Heidegger and Hegel is having conceived of the relationship of Greece and philosophy as an origin and thus as the point of departure of a history internal to the West, such that *philosophy necessarily becomes indistinguishable from its own history*. However close he got to it, Heidegger betrays the movement of deterritorialization because he fixes it once and for all between being and beings, between the Greek territory and the Western earth that the Greeks would have called Being.

Hegel and Heidegger remain historicists inasmuch as they posit history as a form of interiority in which the concept necessarily develops or unveils its destiny. The necessity rests on the abstraction of the historical element rendered circular. The unforeseeable creation of concepts is thus poorly understood. Philosophy is a geophilosophy in precisely the same way that history is a geohistory from Braudel's point of view. Why philosophy in Greece at that moment? It is the same for capitalism, according to Braudel: why capitalism in these places and at these moments? Why not in China at some other moment, since so many of its components were already present there? Geography is not confined to providing historical form with a substance and variable places. It is not merely physical and human but mental, like the landscape. Geography wrests history from the cult of necessity in order to stress the irreducibility of contingency. It wrests it from the cult of origins in order to affirm the power of a "milieu" (what philosophy finds in the Greeks, said Nietzsche, is not an origin but a milieu, an ambiance, an ambient atmosphere: the philosopher ceases to be a comet). It wrests it from structures in order to trace the lines of flight that pass through the Greek world across the Mediterranean. Finally, it wrests history from itself in order to discover becomings that do not belong to history even if they fall back into it: the history of philosophy in Greece must not hide the fact that in every case the Greeks had to become philosophers in the first place, just as philosophers had to become Greek. "Becoming" does not belong to history. History today still designates only the set of conditions, however recent they may be, from which one turns away in order to become, that is to say, in order to create something new. The Greeks did it, but no turning away is valid once and for all. Philosophy cannot be reduced to its own history, because it continually wrests itself from this history in order to create new concepts that fall back into history but do not come from it. How could something come from history? Without history, becoming would remain indeterminate and unconditioned, but becoming is not historical. Psychosocial types belong to history, but conceptual personae belong to becoming. The event itself needs becoming as an unhistorical element. The unhistorical, Nietzsche says, "is like an atmosphere within which alone life can germinate and with the destruction of which it must vanish." It is like a moment of grace; and what "deed would man be capable of if he had not first entered into that vaporous region of the unhistorical?"[7] Philosophy appears in Greece as a result of contingency rather than necessity, as a result of an ambiance or milieu rather than an origin, of a becoming rather than a history, of a geography rather than a historiography, of a grace rather than a nature.

Why did philosophy survive in Greece? We cannot say that capitalism during the Middle Ages is the continuation of the Greek city (even the commercial forms are hardly comparable). But, for always contingent reasons, capitalism leads Europe into a fantastic relative deterritorialization that is due first of all to city-towns *and that itself takes place through immanence.* Territorial produce is connected to an immanent common form able to cross the seas: "wealth in general," "labor *tout court,*" and their coming together as commodity. Marx accurately constructs a concept of capitalism by determining the two principal components, naked labor and pure wealth, with their zone of indiscernibility when wealth buys labor. Why capitalism in the West rather than in China of the third or even the eighth century?[8] Because the West slowly brings together and adjusts these components, whereas the East prevents them from reaching fruition. *Only the West extends and propagates its centers of immanence.* The social field no longer refers to an external limit that restricts it from above, as in the empires, but to immanent internal limits that constantly shift by extending the system, and that reconstitute themselves through displacement.[9] External obstacles are now only technological, and only internal rivalries remain. A world market extends to the ends of the earth before passing into the galaxy: even the skies become horizontal. This is not a result of the Greek endeavor but a resumption, in another form and with other means, on a scale hitherto unknown, which nonetheless relaunches the combination for which the Greeks took the initiative—democratic imperialism, colonizing democracy. The European can, therefore, regard himself, as the Greek did, as not one psychosocial type among others but Man par excellence, and with much more expansive force and missionary zeal than the Greek. Husserl said that, even in their hostility, peoples group themselves into types that have a territorial "home" and family kinship, such as the peoples of India; but only Europe, despite its national rivalries, will propose to itself and other peoples "an incitement to become ever more European," so that in this West the whole of humanity is connected to itself as it never was in Greece.[10] However, it is difficult to believe that it is the rise "of philosophy and the mutually inclusive sciences" that accounts for this privilege of a peculiarly European transcendental subject. Rather, the infinite movement of thought, what Husserl calls Telos, must enter into conjunction with the great relative movement of capital that is continually deterritorialized in order to secure the power of Europe over all other peoples and their reterritorialization on Europe. Modern philosophy's link with capitalism, therefore, is of the same kind as that of ancient philosophy with Greece: *the connection of an absolute plane of immanence with a relative social milieu that also functions through immanence*. From the point of view of philosophy's development, there is no necessary continuity passing from Greece to Europe through the intermediary of Christianity; there is the contingent recommencement of a same contingent process, in different conditions.

The immense relative deterritorialization of world capitalism needs to be reterritorialized on the modern national State, which finds an outcome in democracy, the new society of "brothers," the capitalist version of the society of friends. As Braudel shows, capitalism started out from city-towns, but these pushed deterritorialization so far that immanent modern States had to temper their madness, to recapture and invest them so as to carry out necessary reterritorializations in the form of new internal limits.[11] Capitalism reactivates the Greek world on these economic, political, and social bases. It is the new Athens. The man of capitalism is not Robinson but Ulysses, the cunning plebeian, some average man or other living in the big towns, Autochthonous Proletarians or foreign Migrants who throw themselves into infinite movement—revolution. Not one but two cries traverse capitalism and head for the same disappointment: Immigrants of all countries, unite—workers of all countries. At both ends of the West, America and Russia, pragmatism and socialism play out the return of Ulysses, the new society of brothers or comrades that once again takes up the Greek dream and reconstitutes "democratic dignity."

In fact, the connection of ancient philosophy with the Greek city and the connection of modern philosophy with capitalism are not ideological and do not stop at pushing historical and social determinations to infinity so as to extract spiritual figures from them. Of course, it may be tempting to see philosophy as an agreeable commerce of the mind, which, with the concept, would have its own commodity, or rather its exchange value—which, from the point of view of a lively, disinterested sociability of Western democratic conversation, is able to generate a consensus of opinion and provide communication with an ethic, as art would provide it with an aesthetic. If this is what is called philosophy, it is understandable why marketing appropriates the concept and advertising puts itself forward as the conceiver par excellence, as the poet and thinker. What is most distressing is not this shameless appropriation but the conception of philosophy that made it possible in the first place. The Greeks suffered similar disgraces, relatively speaking, with certain sophists. But what saves modern philosophy is that it is no more the friend of capitalism than ancient philosophy was the friend of the city. Philosophy takes the relative deterritorialization of capital to the absolute; it makes it pass over the plane of immanence as movement of the infinite and suppresses it as internal limit, *turns it back against itself so as to summon forth a new earth, a new people.* But in this way it arrives at the nonpropositional form of the concept in which communication, exchange, consensus, and opinion vanish entirely. It is therefore closer to what Adorno called "negative dialectic" and to what the Frankfurt School called "utopian." Actually, *utopia is what links* philosophy with its own epoch, with European capitalism, but also already with the Greek city. In each case it is with utopia that philosophy becomes political and takes the criticism of its own time to its highest point. Utopia does not split off from infinite movement: etymologically it stands for absolute deterritorialization but always at the critical point at which it is connected with the present relative milieu, and especially with the forces stifled by this milieu. *Erewhon,* the word used by Samuel Butler, refers not only to no-where but also to now-here. What matters is not the supposed distinction between utopian and scientific socialism but the different types of utopia, one of them being revolution. In utopia (as in philosophy) there is always the risk of a restoration, and sometimes a proud affirmation, of transcendence, so that we need to distinguish between authoritarian utopias, or utopias of transcendence, and immanent, revolutionary, libertarian utopias.[12] But to say that revolution is itself utopia of immanence is not to say that it is a dream, something that is not realized or that is only realized by betraying itself. On the contrary, it is to posit revolution as plane of immanence, infinite movement and absolute survey, but to the extent that these features connect up with what is real here and now in the struggle against capitalism, relaunching new struggles whenever the earlier one is betrayed. The word utopia therefore designates *that conjunction of philosophy, or of the concept, with the present milieu*—political philosophy (however, in view of the mutilated meaning public opinion has given to it, perhaps *utopia* is not the best word).

It is not false to say that the revolution "is the fault of philosophers" (although it is not philosophers who lead it). That the two great modern revolutions, American and Soviet, have turned out so badly does not prevent the concept from pursuing its immanent path. As Kant showed, the concept of revolution exists not in the way in which revolution is undertaken in a necessarily relative social field but in the "enthusiasm" with which it is thought on an absolute plane of immanence, like a presentation of the infinite in the here and now, which includes nothing rational or even reasonable.[13] The concept frees immanence from all the limits still imposed on it by capital (or that it imposed on itself in the form of capital appearing as something transcendent). However, it is not so much a case of a separation of the spectator from the actor in this enthusiasm as of a distinction within the action itself between historical factors and "unhistorical vapor," between a state of affairs and the event. As concept and as event, revolution is self-referential or enjoys a self-positing that enables it to be apprehended in an immanent enthusiasm without anything in states of affairs or lived experience being able to tone it down, not even the disappointments of reason. Revolution is absolute deterritorialization even to the point where this calls for a new earth, a new people.

Absolute deterritorialization does not take place without reterritorialization. Philosophy is reterritorialized on the concept. The concept is not object but territory. It does not have an Object but a territory. For that very reason it has a past form, a present form and, perhaps, a form to come. Modern philosophy is reterritorialized on Greece as form of its own past. German philosophers especially have lived the relationship with Greece as a personal relationship. But they indeed lived it as the reverse or contrary of the Greeks, the symmetrical inverse: the Greeks kept the plane of immanence that they constructed in enthusiasm and drunkenness, but they had to search for the concepts with which to fill it so as to avoid falling back into the figures of the East. As for us, we possess concepts—after so many centuries of Western thought we think we possess them—but we hardly know where to put them because we lack a genuine plane, misled as we are by Christian transcendence. In short, in its past form the concept is that which was not yet. We today possess concepts, but the Greeks did not yet possess them; they possessed the plane that we no longer possess. That is why Plato's Greeks *contemplate* the concept as something that is still very far away and beyond, whereas we possess the concept—we possess it in the mind innately; all that is needed is to *reflect*. This is what Hölderlin expressed so profoundly: the "Autochthon" for the Greeks is our "stranger," that which we have to acquire, whereas our Autochthon is what, to the contrary, the Greeks had to acquire as their stranger.[14] Or, as Schelling put it, the Greeks lived and thought in Nature but left Mind in the "mysteries," whereas we live, think, and feel in the Mind, in reflection, but leave Nature in a profound alchemical mystery that we constantly profane. The Autochthon and the stranger are no longer separate, like two distinct personae, but distributed like one and the same double persona who unfolds into two versions in turn, present and past: what was Autochthonous becomes strange; what was strange becomes Autochthonous. With all his strength Hölderlin calls for a "society of friends" as the condition of thought, but it is as if this society had suffered a catastrophe that changes the nature of friendship. We reterritorialize ourselves among the Greeks but according to what they did not possess and had not yet become, so that we reterritorialize them on ourselves.

Philosophical reterritorialization therefore also has a present form. Can we say that philosophy is reterritorialized on the modern democratic State and human rights? But because there is no universal democratic State this movement implies the particularity of a State, of a right, or of the spirit of a people capable of expressing human rights in "its" State and of outlining the modern society of brothers. In fact, it is not only the philosopher, as man, who has a nation; it is philosophy that is reterritorialized on the national State and the spirit of the people (usually those of the philosopher, but not always). Thus Nietzsche founded geophilosophy by seeking to determine the national characteristics of French, English, and German philosophy. But why were only three countries collectively able to produce philosophy in the capitalist world? Why not Spain or Italy? Italy in particular presented a set of deterritorialized cities and a maritime power that were capable of reviving the conditions for a "miracle." It marked the start of an incomparable philosophy. But it aborted, with its heritage passing instead to Germany (with Leibniz and Schelling). Perhaps Spain was too subject to the Church and Italy too "close" to the Holy See. Perhaps it was the break with Catholicism that saved England and Germany spiritually, and perhaps Gallicanism* was what saved France. Italy and Spain lacked a "milieu" for philosophy, so that their thinkers remained "comets"; and they were inclined to burn their comets. Italy and Spain were the two Western countries capable of a powerful development of *concettism,* that is to say, of that Catholic compromise of concept and figure which had great aesthetic value but which masked philosophy, diverted it toward a rhetoric and prevented a full possession of the concept.

The present form is expressed thus: we have concepts! The Greeks, however, did not yet "have" them and contemplated them from afar, or sensed them: the difference between Platonic reminiscence and Cartesian innateness or the Kantian a priori derives from this. But possession of the concept does not appear to coincide with revolution, the democratic State, and human rights. If in America the philosophical enterprise of pragmatism, so poorly understood in France, has continuities with the democratic revolution and the new society of brothers, this is not true of the golden age of seventeenth-century French philosophy, or of eighteenth-century England, or of nineteenth-century Germany. But this is only to say that human history and the history of philosophy do not have the same rhythm. French philosophy already speaks in the name of a republic of minds and of a capacity to think as something that is "the most widely shared" and that will end up being expressed in a revolutionary cogito. England will constantly reflect on its revolutionary experience and will be the first to ask why revolutions turn out so badly in reality when in spirit they promise so much. England, America, and France exist as the three lands of human rights. As for Germany, it will continue to reflect on the French revolution from its side, as that which it cannot do (it lacks sufficiently deterritorialized towns; it suffers from the weight of a hinterland, the *Land*). But what it cannot

---

*"Gallicanism" refers to the movement within the French Catholic Church that sought to maintain the distinctive characteristics of the Church in Gaul or France and asserted the right of the French Church to a certain degree of independence from Rome.

do it undertakes to think. In each case philosophy finds a way of reterritorializing itself in the modern world in conformity with the spirit of a people and its conception of right. The history of philosophy therefore is marked by national characteristics or rather by nationalitarianisms [*nationalitaires**], which are like philosophical "opinions."

EXAMPLE 8

If we moderns possess the concept but have lost sight of the plane of immanence, then the tendency of the French persona in philosophy is to manage this situation by supporting concepts through a simple order of reflexive knowledge, an order of reasons, an "epistemology." It is like the inventory of habitable, civilizable, knowable or known lands that are summed up by an awareness or cogito, even if this cogito must become prereflexive, and this consciousness must become nonthetic, so as to cultivate what is most barren. The French are like landowners whose source of income is the cogito. They are always reterritorialized on consciousness. Germany, on the other hand, does not give up the absolute: it makes use of consciousness but as a means of deterritorialization. It wants to reconquer the Greek plane of immanence, the unknown earth that it now feels as its own *barbarism,* its own *anarchy* abandoned to the *nomads* since the disappearance of the Greeks.[15] It must also constantly clear and consolidate this ground, that is to say, it must lay foundations. A mania for founding, for conquering, inspires this philosophy; what the Greeks possessed Autochthonously, German philosophy would have through conquest and foundation, so that it would make immanence immanent *to* something, to

---

*We translate *nationalitaire* as "nationalitarian," in line with the translation of *totalitaire* as "totalitarian."

its own Act of philosophizing, to its own philosophizing subjectivity (the cogito therefore takes on a different meaning since it conquers and lays down the ground).

England, from this point of view, is Germany's obsession, for the English are precisely those nomads who treat the plane of immanence as a movable and moving ground, a field of radical experience, an archipelagian world where they are happy to pitch their tents from island to island and over the sea. The English nomadize over the old Greek earth, broken up, fractalized, and extended to the entire universe. We cannot even say that they have concepts like the French and Germans; but they acquire them, they only believe in what is acquired—not because everything comes from the senses but because a concept is acquired by inhabiting, by pitching one's tent, by contracting a habit. In the trinity Founding-Building-Inhabiting, the French build and the Germans lay foundations, but the English inhabit. For them a tent is all that is needed. They develop an extraordinary conception of habit: habits are taken on by contemplating and by contracting that which is contemplated. Habit is creative. The plant contemplates water, earth, nitrogen, carbon, chlorides, and sulphates, and it contracts them in order to acquire its own concept and fill itself with it (enjoyment*). The concept is a habit acquired by contemplating the elements from which we come (hence the very special Greekness of English philosophy, its empirical neo-Platonism). We are all contemplations, and therefore habits. *I* is a habit. Wherever there are habits there are concepts, and habits are developed and given up on the plane of immanence of radical experience: they are "conventions."[16] That is why English philosophy is a free and wild creation of concepts. To what convention is a

---

*In English in the original.

given proposition due; what is the habit that constitutes its concept? This is the question posed by pragmatism. English law is a law of custom and convention, as the French is of contract (deductive system) and the German of institution (organic totality). When philosophy is reterritorialized on the State of Law, the philosopher becomes philosophy professor; but for the German this is by institution and foundation, for the French it is by contract, and for the English it is solely by convention.

If there is no universal democratic State, despite German philosophy's dream of foundation, it is because the market is the only thing that is universal in capitalism. In contrast with the ancient empires that carried out transcendent overcodings, capitalism functions as an immanent axiomatic of decoded flows (of money, labor, products). National States are no longer paradigms of overcoding but constitute the "models of realization" of this immanent axiomatic. In an axiomatic, models do not refer back to a transcendence; quite the contrary. It is as if the deterritorialization of States tempered that of capital and provided it with compensatory reterritorializations. Now, models of realization may be very diverse (democratic, dictatorial, totalitarian), they may be really heterogeneous, but they are nonetheless isomorphous with regard to the world market insofar as the latter not only presupposes but produces determinate inequalities of development. That is why, as has often been noted, democratic States are so bound up with, and compromised by, dictatorial States that the defense of human rights must necessarily take up the internal criticism of every democracy. Every democrat is also the "other Tartuffe" of Beaumarchais, the humanitarian Tartuffe, as Péguy said. Of course, there is no reason to believe that we can no longer think after Auschwitz, or that we are all responsible for Nazism in an unwholesome culpability that, moreover, would only affect the victims. As Primo Levi said, they will not make us confuse the victims with the executioners. But, he says, what Nazism and the camps inspire in us is much more or much less: "the shame of being a man" (because even the survivors had to collude, to compromise themselves).[17] It is not only our States but each of us, every democrat, who finds him or herself not responsible for Nazism but sullied by it. There is indeed catastrophe, but it consists in the society of brothers or friends having undergone such an ordeal that brothers and friends can no longer look at each other, or each at himself, without a "weariness," perhaps a "mistrust," which does not suppress friendship but gives it its modern color and replaces the simple "rivalry" of the Greeks. We are no longer Greeks, and friendship is no longer the same: Blanchot and Mascolo have seen the importance of this mutation for thought itself.

Human rights are axioms. They can coexist on the market with many other axioms, notably those concerning the security of property, which are unaware of or suspend them even more than they contradict them: "the impure mixture or the impure side by side," said Nietzsche. Who but the police and armed forces that coexist with democracies can control and manage poverty and the deterritorialization-reterritorialization of shanty towns? What social democracy has not given the order to fire when the poor come out of their territory or ghetto? Rights save neither men nor a philosophy that is reterritorialized on the democratic State. Human rights will not make us bless capitalism. A great deal of innocence or cunning is needed by a philosophy of communication that claims to restore the society of friends, or even of wise men, by forming a universal opinion as "consensus" able to moralize nations, States, and the market.[18] Human rights say nothing about the immanent modes of existence of people provided with rights. Nor is it only in the extreme situations described by Primo Levi that we experience the shame of being human. We also experience it in insignificant conditions, before the meanness and vulgarity of existence that haunts democracies, before the propagation of these modes of existence and of thought-for-the-market, and before the values, ideals, and opinions of our time. The ignominy of the possibilities of life that we are offered appears from within. We do not feel ourselves outside of our time but continue to undergo shameful compromises with it. This feeling of shame is one of philosophy's most powerful motifs. We are not responsible for the victims but responsible before them. And there is no way to escape the ignoble but to play the part of the animal (to growl, burrow, snigger, distort ourselves): thought itself is sometimes closer to an animal that dies than to a living, even democratic, human being.

If philosophy is reterritorialized on the concept, it does not find the condition for this in the present form of the democratic State or in a cogito of communication that is even more dubious than that of reflection. We do not lack communication. On the contrary, we have too much of it. We lack creation. *We lack resistance to the present.* The creation of concepts in itself calls for a future form, for a new earth and people that do not yet exist. Europeanization does not constitute a becoming but merely the history of capitalism, which prevents the becoming of subjected peoples. Art and philosophy converge at this point: the constitution of an earth and a people that are lacking as the correlate of creation. It is not populist writers but the most aristocratic who lay claim to this future. This people and earth will not be found in our democracies. Democracies are majorities, but a becoming is by its nature that which always eludes the majority. The position of many writers with respect to democracy is complex and ambiguous. The Heidegger affair has complicated matters: a great philosopher actually had to be reterritorialized on Nazism for the strangest commentaries to meet up, sometimes calling his philosophy into question and sometimes absolving it through such complicated and convoluted arguments that we are still in the dark. It is not always easy to be Heideggerian. It would be easier to understand a great painter or musician falling into shame in this way (but, precisely, they did not). It had to be a philosopher, as if shame had to enter into philosophy itself. He wanted to rejoin the Greeks through the Germans, at the worst moment in their history: is there anything worse, said Nietzsche, than to find oneself facing a German when one was expecting a Greek? How could Heidegger's concepts not be intrinsically sullied by an abject reterritorialization? Unless all concepts include this gray zone and indiscernibility where for a moment the combatants on the ground are confused, and the thinker's tired eye mistakes one for the other—not only the German for a Greek but the fascist for a creator of existence and freedom. Heidegger lost his way along the paths of the reterritorialization because they are paths without directive signs or barriers. Perhaps this strict professor was madder than he seemed. He got the wrong people, earth, and blood. For the race summoned forth by art or philosophy is not the one that claims to be pure but rather an oppressed, bastard, lower, anarchical, nomadic, and irremediably minor race—the very ones that Kant excluded from the paths of the new Critique. Artaud said: to write *for* the illiterate—to speak for the aphasic, to think for the acephalous. But what does "for" mean? It is not "for their benefit," or yet "in their place." It is "before." It is a question of becoming. The thinker is not acephalic, aphasic, or illiterate, but becomes so. He becomes Indian, and never stops becoming so—perhaps "so that" the Indian who is himself Indian becomes something else and tears himself away from his own agony. We think and write for animals themselves. We become animal so that the animal also becomes something else. The agony of a rat or the slaughter of a calf remains present in thought not through pity but as the zone of exchange between man and animal in which something of one passes into the other. This is the constitutive relationship of philosophy with nonphilosophy. Becoming is always double, and it is this double becoming that constitutes the people to come and the new earth. The philosopher must become nonphilosopher so that nonphilosophy becomes the earth and people of philosophy. Even such a well-respected philosopher as Bishop Berkeley never stops saying, "We Irish others, the mob." The people is internal to the thinker because it is a "becoming-people," just as the thinker is internal to the people as no less unlimited becoming.

The artist or the philosopher is quite incapable of creating a people, each can only summon it with all his strength. A people can only be created in abominable sufferings, and it cannot be concerned any more with art or philosophy. But books of philosophy and works of art also contain their sum of unimaginable sufferings that forewarn of the advent of a people. They have resistance in common—their resistance to death, to servitude, to the intolerable, to shame, and to the present.

Deterritorialization and reterritorialization meet in the double becoming. The Autochthon can hardly be distinguished from the stranger because the stranger becomes Autochthonous in the country of the other who is not, at the same time that the Autochthon becomes stranger to himself, his class, his nation, and his language: we speak the same language, and yet I do not understand you. Becoming stranger to oneself, to one's language and nation, is not this the peculiarity of the philosopher and philosophy, their "style," or what is called a philosophical gobbledygook? *In short, philosophy is reterritorialized three times:* on the Greeks in the past, on the democratic State in the present, and on the new people and earth in the future. Greeks and democrats are strangely deformed in this mirror of the future.

Utopia is not a good concept because even when opposed to History it is still subject to it and lodged within it as an ideal or motivation. But becoming is the concept itself. It is born in History, and falls back into it, but is not of it. In itself it has neither beginning nor end but only a milieu. It is thus more geographical than historical. Such are revolutions and societies of friends, societies of resistance, because to create is to resist: pure becomings, pure events on a plane of immanence. What History grasps of the event is its effectuation in states of affairs or in lived experience, but the event in its becoming, in its specific consistency, in its self-positing as concept, escapes History. Psychosocial types are historical, but conceptual personae are events. Sometimes one ages in accordance with History, and with it, sometimes one becomes old in a quite unobtrusive event (perhaps the same event that allows the problem "what is philosophy?" to be posed). And it is the same for those who die young—there are several ways of so dying. To think is to experiment, but experimentation is always that which is in the process of coming about—the new, remarkable, and interesting that replace the appearance of truth and are more demanding than it is. What is in the process of coming about is no more what ends than what begins. History is not experimentation, it is only the set of almost negative conditions that make possible the experimentation of something that escapes history. Without history experimentation would remain indeterminate and unconditioned, but experimentation is not historical. It is philosophical.

EXAMPLE 9

In a great work of philosophy, Péguy explains that there are two ways of considering the event. One consists in going over the course of the event, in recording its effectuation in history, its conditioning and deterioration in history. But the other consists in reassembling the event, installing oneself in it as in a becoming, becoming young again and aging in it, both at the same time, going through all its components or singularities. It may be that nothing changes or seems to change in history, but everything changes, and we change, in the event: "There was nothing. Then a problem to which we saw no end, a problem without solution . . . suddenly no longer exists and we wonder what we were talking about"; it has gone into other problems; "there was nothing and one is in a new people, in a new world, in a new man."[19] This is no longer the historical, and it is not the eternal, Péguy says: it is the *Aternal* [*Internel*]. Péguy had to create this noun to designate a new concept. Is this not something similar to that which a thinker far from Péguy designated *Untimely* or

Inactual—the unhistorical vapor that has nothing to do with the eternal, the becoming without which nothing would come about in history but that does not merge with history? Beneath the Greeks and States, it launches a people, an earth, like the arrow and discus of a new world that is never-ending, that is always in the process of coming about— "acting counter to time, and therefore acting on our time and, let us hope, for the benefit of a time to come." Acting counter to the past, and therefore on the present, for the benefit, let us hope, of a future—but the future is not a historical future, not even a utopian history, it is the infinite Now, the *Nun* that Plato already distinguished from every present: the Intensive or Untimely, not an instant but a becoming. Again, is this not what Foucault called the *Actual?* But how could the concept now be called the actual when Nietzsche called it the inactual? Because, for Foucault, what matters is the difference between the present and the actual. The actual is not what we are but, rather, what we become, what we are in the process of becoming—that is to say, the Other, our becoming-other. The present, on the contrary, is what we are and, thereby, what already we are ceasing to be. We must distinguish not only the share that belongs to the past and the one that belongs to the present but, more profoundly, the share that belongs to the present and that belonging to the actual.[20] It is not that the actual is the utopian prefiguration of a future that is still part of our history. Rather, it is the now of our becoming. When Foucault admires Kant for posing the problem of philosophy in relation not to the eternal but to the Now, he means that the object of philosophy is not to contemplate the eternal or to reflect history but to diagnose our actual becomings: a becoming-revolutionary that, according to Kant himself, is not the same thing as the past, present, or future of revolutions. A becoming-democratic that is not the same as what States of law are, or even a becoming-Greek that is not the same as what the Greeks were. The *diagnosis* of becomings in every passing present is what Nietzsche assigned to the philosopher as physician, "physician of civilization," or inventor of new immanent modes of existence. Eternal philosophy, but also the history of philosophy, gives way to a becoming-philosophical. What becomings pass through us today, which sink back into history but do not arise from it, or rather that arise from it only to leave it? The Aternal, the Untimely, the Actual are examples of concepts in philosophy; exemplary concepts. And if one calls Actual what the other called Inactual, this is only in virtue of a combination of the concept, in virtue of its proximities and components, the slight displacements of which entail, as Péguy said, the modification of a problem (the Temporally eternal in Péguy, the Eternity of becoming according to Nietzsche, and the Outside-interior with Foucault).

Part Two

Philosophy, Science, Logic, and Art

5. Functives and Concepts

The object of science is not concepts but rather functions that are presented as propositions in discursive systems. The elements of functions are called *functives*. A scientific notion is defined not by concepts but by functions or propositions. This is a very complex idea with many aspects, as can be seen already from the use to which it is put by mathematics and biology respectively. Nevertheless, it is this idea of the function which enables the sciences to reflect and communicate. Science does not need philosophy for these tasks. On the other hand, when an object—a geometrical space, for example—is scientifically constructed by functions, its philosophical concept, which is by no means given in the function, must still be discovered. Furthermore, a concept may take as its components the functives of any possible function without thereby having the least scientific value, but with the aim of marking the differences in kind between concepts and functions.

Under these conditions, the first difference between science and philosophy is their respective attitudes toward chaos. Chaos is defined not so much by its disorder as by the infinite speed with which every form taking shape in it vanishes. It is a void that is not a nothingness but a *virtual,* containing all possible particles and drawing out all possible forms, which spring up only to disappear immediately, without consistency or reference, without consequence.[1] Chaos is an infinite speed of birth and disappearance. Now philosophy wants to know how to retain infinite speeds while gaining consistency, by *giving the virtual a consistency specific to it.* The philosophical sieve, as plane of immanence that cuts through the chaos, selects infinite movements of thought and is filled with concepts formed like consistent particles going as fast as thought. Science approaches chaos in a completely different, almost opposite way: it relinquishes the infinite, infinite speed, in order to gain *a reference able to actualize the virtual.* By retaining the infinite, philosophy gives consistency to the virtual through concepts; by relinquishing the infinite, science gives a reference to the virtual, which actualizes it through functions. Philosophy proceeds with a plane of immanence or consistency; science with a plane of reference. In the case of science it is like a freeze-frame. It is a fantastic *slowing down,* and it is by slowing down that matter, as well as the scientific thought able to penetrate it with propositions, is actualized. A function is a Slow-motion. Of course, science constantly advances accelerations, not only in catalysis but in particle accelerators and expansions that move galaxies apart. However, the primordial slowing down is not for these phenomena a zero-instant with which they break but rather a condition coextensive with their whole development. To slow down is to set a limit in chaos to which all speeds are subject, so that they form a variable determined as abscissa, at the same time as the limit forms a universal constant that cannot be gone beyond (for example, a maximum degree of contraction). The first functives are therefore the limit and the variable, and reference is a relationship between values of the variable or, more profoundly, the relationship of the variable, as abscissa of speeds, with the limit.

Sometimes the constant-limit itself appears as a relationship in the whole of the universe to which all the parts are subject under a finite condition (quantity of movement, force, energy). Again, there must be systems of coordinates to which the terms of the relationship refer: this, then, is a second sense of limit, an external framing or exoreference. For these protolimits, outside all coordinates, initially generate speed abscissas on which axes will be set up that can be coordinated. A particle will have a position, an energy, a mass, and a spin value but on condition that it receives a physical existence or actuality, or that it "touches down" in trajectories that can be grasped by systems of coordinates. It is these first limits that constitute slowing down in the chaos or the threshold of suspension of the infinite, which serve as endoreference and carry out a counting: they are not relations but numbers, and the entire theory of functions depends on numbers. We refer to the speed of light, absolute zero, the quantum of action, the Big Bang: the absolute zero of temperature is minus 273.15 degrees Centigrade, the speed of light, 299,796 kilometers per second, where lengths contract to zero and clocks stop. Such limits do not apply through the empirical value that they take on solely within systems of coordinates, they act primarily as the condition of primordial slowing down that, in relation to infinity, extends over the whole scale of corresponding speeds, over their conditioned accelerations or slowing-downs. It is not only the diversity of these limits that entitles us to doubt the unitary vocation of science. In fact, each limit on its own account generates irreducible, heterogeneous systems of coordinates and imposes thresholds of discontinuity depending on the proximity or distance of the variable (for example, the distance of the galaxies). Science is haunted not by its own unity but by the plane of reference constituted by all the limits or borders through which it confronts chaos. It is these borders that give the plane its references. As for the systems of coordinates, they populate or fill out the plane of reference itself.

EXAMPLE 10

It is difficult to see how the limit immediately cuts into the infinite, the unlimited. Yet it is not the limited thing that sets a limit to the infinite but the limit that makes possible a limited thing. Pythagoras, Anaximander, and Plato himself understood this: the limit and the infinite clasped together in an embrace from which things will come. Every limit is illusory and every determination is negation, if determination is not in an immediate relation with the undetermined. The theory of science and of functions depends on this. Later, Cantor provides this theory with its mathematical formulas from a double—intrinsic and extrinsic—point of view. According to the first, a set is said to be infinite if it presents a term-by-term correspondence with one of its parts or subsets, the set and the subset having the same power or the same number of elements that can be designated by "aleph 0," as with the set of whole numbers. According to the second determination, the set of subsets of a given set is necessarily larger than the original set: the set of aleph 0 subsets therefore refers to a different transfinite number, aleph 1, which possesses the power of the continuum or corresponds to the set of real numbers (we then continue with aleph 2, etc.). It is odd that this conception has so often been seen as reintroducing infinity into mathematics: it is, rather, the extreme consequence of the definition of the limit by a number, this being the first whole number that follows all the finite whole numbers none of which is maximum. What the theory of sets does is inscribe the limit within the infinite itself, without which there could be no limit: in its strict hierarchization it installs a slowing-down, or rather, as Cantor himself says, a stop—a "principle of stopping" whereby a new whole number is created only "if the rounding up of all the preceding numbers has the power of a class of definite numbers, already given in its whole extension."[2] Without this principle of stopping or of slowing down, there would be a set of all sets that Cantor already rejects and which, as Russell demonstrates, could only be chaos. Set theory is the constitution of a plane of reference, which includes not only an *endoreference* (intrinsic determination of an infinite set) but also an *exoreference* (extrinsic determination). In spite of the explicit attempt by Cantor to unite philosophical concept and scientific function, the characteristic difference remains, since the former unfolds on a plane of immanence or consistency without reference, but the other on a plane of reference devoid of consistency (Gödel).

When the limit generates an abscissa of speeds by slowing down, the virtual forms of chaos tend to be actualized in accordance with an ordinate. And certainly the plane of reference already carries out a preselection that matches forms to the limits or even to the regions of particular abscissas. But the forms nonetheless constitute variables independent of those that move by abscissa. This is very different from the philosophical concept: intensive ordinates no longer designate inseparable components condensed in the concept as absolute survey (variations) but rather distinct determinations that must be matched in a discursive formation with other determinations taken in extension (variables). Intensive ordinates of forms must be coordinated with extensive abscissas of speed in such a way that speeds of development and the actualization of forms relate to each other as distinct, extrinsic determinations.[3] It is from this second point of view that the limit is now the origin of a system of coordinates made up of at least two independent variables; but these enter into a relation on which a third variable depends as state of affairs or formed matter in the system (such states of affairs may be mathematical, physical, biological). This is indeed the new meaning of reference as form of the proposition, the relation of a state of affairs to the system. The state of affairs is a function: it is a complex variable that depends on a relation between at least two independent variables.

The respective independence of variables appears in mathematics when one of them is at a higher power than the first. That is why Hegel shows that variability in the function is not confined to values that can be changed ($^2/_3$ and $^4/_6$) or are left undetermined (a = 2b) but requires one of the variables to be at a higher power (y^2/x = P). For it is then that a relation can be directly determined as differential relation dy/dx, in which the only determination of the value of the variables is that of disappearing or being born, even though it is wrested from infinite speeds. A state of affairs or "derivative" function depends on such a relation: an operation of depotentialization has been carried out that makes possible the comparison of distinct powers starting from which a thing or a body may well develop (integration).[4] In general, a state of affairs does not actualize a chaotic virtual without taking from it a *potential* that is distributed in the system of coordinates. From the virtual that it actualizes it draws a potential that it appropriates. The most closed system still has a thread that rises toward the virtual, and down which the spider descends. But knowing whether the potential can be re-created in the actual, whether it can be renewed and enlarged, allows us to distinguish states of affairs, things, and bodies more precisely. When we go from the state of affairs to the *thing* itself, we see that a thing is always related to several axes at once according to variables that are functions of each other, even if the internal unity remains undetermined. But, when the thing itself undergoes changes of coordinates, strictly speaking it becomes a *body,* and instead of the function taking the limit and the variable as reference, it takes an invariant and a group of transformations (the Euclidean body of geometry, for example, is constituted by invariants in relation to the group of movements). The "body," in fact, is not here the special field of biology, and it finds a mathematical determination on the basis of an absolute minimum represented by the rational numbers by carrying out independent extensions of this basic body that increasingly limit possible substitutions until there is a perfect individuation. The difference between body and state of affairs (or thing) pertains to the individuation of the body, which proceeds by a cascade of actualizations. With bodies, the relationship between independent variables becomes fully worked out, even if it means providing itself with a potential or power that renews its individuation. Particularly when the body is a living being, which proceeds by differentiation and no longer by extension or addition, a new type of variable arises, internal variables determining specifically biological functions in relation to internal milieus (endoreference) but also entering into probabilistic functions with external variables of the outside milieu (exoreference).[5]

Thus we find ourselves confronting a new string of functives, systems of coordinates, potentials, states of affairs, things, and bodies. States of affairs are ordered mixtures, of very different types, which may even only concern trajectories. But things are interactions, and bodies are communications. States of affairs refer to geometrical coordinates of supposedly closed systems, things refer to energetic coordinates of coupled systems, and bodies refer to the informational coordinates of separated, unconnected systems. The history of the sciences is inseparable from the construction, nature, dimensions, and proliferation of axes. Science does not carry out any unification of the Referent but produces all kinds of bifurcations on a plane of reference that does not preexist its detours or its layout. It is as if the bifurcation were searching the infinite chaos of the virtual for new forms to actualize by carrying out a sort of potentialization of matter: carbon introduces a bifurcation into Mendeleyev's table, which, through its plastic properties, produces the state of organic matter. The problem of a unity or multiplicity of science, therefore, must not be posed as a function of a system of coordinates that is possibly unique at a given moment. As with the plane of immanence in philosophy, we must ask what status before and after assume, simultaneously, on a plane of reference with temporal dimension and evolution. Is there just one or several planes of reference? The answer will not be the same as the one given for the philosophical plane of immanence with its strata or superimposed layers. This is because reference, implying a renunciation of the infinite, can only connect up chains of functives that necessarily break at some point. The bifurcations, slowing-downs, and accelerations produce holes, breaks, and ruptures that refer back to other variables, other relations, and other references. According to some basic examples, it is said that the fractional number breaks with the whole number, irrational with rational numbers, Riemannian with Euclidean geometry. But in the other simultaneous direction, from after to before, the whole number appears as a particular case of the fractional number, or the rational as a particular case of a "break" in a linear set of points. It is true that this unifying process that works in the retroactive direction necessarily brings in other references, the variables of which are subject not only to restrictive conditions for giving the particular case but, in themselves, to new ruptures and bifurcations that will change their own references. This is what happens when Newton is derived from Einstein, or real numbers from the break, or Euclidean geometry from an abstract metrical geometry—which amounts to saying with Kuhn that science is *paradigmatic,* whereas philosophy is syntagmatic.

Science is not confined to a linear temporal succession any more than philosophy is. But, instead of a stratigraphic time, which expresses before and after in an order of superimpositions, science displays a peculiarly serial, ramified time, in which the before (the previous) always designates bifurcations and ruptures to come, and the after designates retroactive reconnections. This results in a completely different pace of scientific progress. Scientists' proper names are written in this other time, this other element, marking points of rupture and points of reconnection. Of course, it is always possible, and sometimes fruitful, to interpret the history of philosophy according to this scientific rhythm. But to say that Kant breaks with Descartes, and that the Cartesian cogito becomes a particular case of the Kantian cogito, is not entirely satisfying since this is, precisely, to turn philosophy into a science (conversely, it would be no more satisfying to establish an order of superimposition between Newton and Einstein). Far from forcing us to pass through the same components again, the function of the scientist's proper name is to spare us from doing this and to persuade us that there is no reason to go down the same path again: we do not work through a named equation, we use it. Far from distributing cardinal points that organize syntagms on a plane of immanence, the scientist's proper name draws up paradigms that are projected into necessarily oriented systems of reference. Finally, the relationship of science with philosophy is less of a problem than that of its even more passionate relationship with religion, as can be seen in all the attempts at scientific uniformization and universalization in the search for a single law, a single force, or a single interaction. What brings science and religion together is that functives are not concepts but figures defined by a spiritual tension rather than by a spatial intuition. There is something figural in functives that forms an *ideography* peculiar to science and that already makes vision a reading. But what constantly reaffirms the opposition of science to all religion and, at the same time, happily makes the unification of science impossible is the substitution of reference for all transcendence. It is the functional correspondence of the paradigm with a system of reference that, by determining an exclusively scientific way in which the figure must be *constructed, seen, and read* through functives, prohibits any infinite religious utilization of the figure.[6]

The first difference between philosophy and science lies in the respective presuppositions of the concept and the function: in the one a plane of immanence or consistency, in the other a plane of reference.

The plane of reference is both one and multiple, but in a different way from the plane of immanence. The second difference concerns the concept and the function more directly: the inseparability of variations is the distinctive characteristic of the unconditioned concept, while the independence of variables, in relationships that can be conditioned, is essential to the function. In one case we have a set of *inseparable variations* subject to "a contingent reason" that constitutes the concept from variations; and in the other case we have a set of *independent variables* subject to "a necessary reason" that constitutes the function from variables. That is why, from this point of view, the theory of functions presents two poles depending on whether, n variables being given, one can be considered as function of the $n - 1$ independent variables, with $n - 1$ partial derivatives and a differential total of the function, or, on the contrary, whether $n - 1$ magnitudes are functions of a single independent variable, without differential total of the composite function. In the same way, the problem of tangents (differentiation) summons as many variables as there are curves in which the derivative for each is any tangent whatever at any point whatever. But the inverse problem of tangents (integration) deals with only a single variable, which is the curve itself tangent to all the curves of the same order, on condition of a change of coordinates.[7] An analogous duality concerns the dynamic description of a system of n independent particles: the instantaneous state can be represented by n points and n vectors of speed in a three-dimensional space but also by a single point in a phase space.

It could be said that science and philosophy take opposed paths, because philosophical concepts have events for consistency whereas scientific functions have states of affairs or mixtures for reference: through concepts, philosophy continually extracts a consistent event from the state of affairs—a smile without the cat, as it were—whereas through functions, science continually actualizes the event in a state of affairs, thing, or body that can be referred to. From this point of view, the pre-Socratics had already grasped the essential point for a determination of science, valid right up to our own time, when they made physics a theory of mixtures and their different types.[8] And the Stoics carried to its highest point the fundamental distinction between, on the one hand, states of affairs or mixtures of bodies in which the event is actualized and, on the other, incorporeal events that rise like a vapor from states of affairs themselves. It is, therefore, through two linked characteristics that philosophical concept and scientific function are distinguished: inseparable variations and independent variables; events on a plane of immanence and states of affairs in a system of reference (the different status of intensive ordinates in each case derives from this since they are internal components of the concept, but only coordinates of extensive abscissas in functions, when variation is no more than a state of variable). *Concepts and functions thus appear as two types of multiplicities or varieties whose natures are different.* Although scientific types of multiplicity are themselves extremely diverse, they do not include the properly philosophical multiplicities for which Bergson claimed a particular status defined by duration, "multiplicity of fusion," which expressed the inseparability of variations, in contrast to multiplicities of space, number, and time, which ordered mixtures and referred to the variable or to independent variables.[9] It is true that this very opposition, between scientific and philosophical, discursive and intuitive, and extensional and intensive multiplicities, is also appropriate for judging the correspondence between science and philosophy, their possible collaboration, and the inspiration of one by the other.

Finally, there is a third major difference, which no longer concerns the respective presuppositions or the element as concept or function but the *mode of enunciation.* To be sure, there is as much experimentation in the form of thought experiment in philosophy as there is in science, and, being close to chaos, the experience can be overwhelming in both. But there is also as much creation in science as there is in philosophy or the arts. There is no creation without experiment. Whatever the difference between scientific and philosophical lan-

guages and their relationship with so-called natural languages, functives (including axes of coordinates) do not preexist ready-made any
more than concepts do. Granger has shown that in scientific systems
"styles" associated with proper names have existed—not as an extrinsic determination but, at the least, as a dimension of their creation
and in contact with an experience or a lived[10] [*un vécu*]. Coordinates,
functions and equations, laws, phenomena or effects, remain attached
to proper names, just as an illness is called by the name of the
physician who succeeded in isolating, putting together, and cluster-
ing its variable signs. Seeing, seeing what happens, has always had a
more essential importance than demonstrations, even in pure mathematics, which can be called visual, figural, independently of its applications: many mathematicians nowadays think that a computer is
more precious than an axiomatic, and the study of nonlinear functions
passes through slownesses and accelerations in series of observable
numbers. The fact that science is discursive in no way means that it
is deductive. On the contrary, in its bifurcations it undergoes many
catastrophes, ruptures, and reconnections marked by proper names.
If there is a difference between science and philosophy that is impossible to overcome, it is because proper names mark in one case a
juxtaposition of reference and in the other a superimposition of layer:
they are opposed to each other through all the characteristics of
reference and consistency. But on both sides, philosophy and science
(like art itself with its third side) include an *I do not know* that has
become positive and creative, the condition of creation itself, and that
consists in determining *by* what one does not know—as Galois said,
"indicating the course of calculations and anticipating the results
without ever being able to bring them about."[11]

   We are referred back to another aspect of enunciation that applies
no longer to proper names of scientists or philosophers but to their
ideal intercessors internal to the domains under consideration. We
saw earlier the philosophical role of *conceptual personae* in relation to
fragmentary concepts on a plane of immanence, but now science

brings to light *partial observers* in relation to functions within systems of reference. The fact that there is no total observer that, like Laplace's "demon," is able to calculate the future and the past starting from a given state of affairs means only that God is no more a scientific observer than he is a philosophical persona. But "demon" is still excellent as a name for indicating, in philosophy as well as in science, not something that exceeds our possibilities but a common kind of these necessary intercessors as respective "subjects" of enunciation: the philosophical friend, the rival, the idiot, the overman are no less demons than Maxwell's demon or than Einstein's or Heisenberg's observers. It is not a question of what they can or cannot do but of the way in which they are perfectly positive, from the point of view of concept or function, even in what they do not know and cannot do. In both cases there is immense variety, but not to the extent of forgetting the different natures of the two great types.

To understand the nature of these partial observers that swarm through all the sciences and systems of reference, we must avoid giving them the role of a limit of knowledge or of an enunciative subjectivity. It has been noted that Cartesian coordinates privilege the points situated close to the origin, whereas those of projective geometry gave "a finite image of all the values of the variable and the function." But perspective fixes a partial observer, like an eye, at the summit of a cone and so grasps contours without grasping reliefs or the quality of the surface that refer to another observer position. As a general rule, the observer is neither inadequate nor subjective: even in quantum physics, Heisenberg's demon does not express the impossibility of measuring both the speed and the position of a particle on the grounds of a subjective interference of the measure with the measured, but it measures exactly an objective state of affairs that leaves the respective position of two of its particles outside of the field of its actualization, the number of independent variables being reduced and the values of the coordinates having the same probability. Subjectivist interpretations of thermodynamics, relativity, and quantum physics manifest the same inadequacies. Perspectivism, or scientific relativism, is never relative to a subject: it constitutes not a relativity of truth but, on the contrary, a truth of the relative, that is to say, of variables whose cases it orders according to the values it extracts from them in its system of coordinates (here the order of conic sections is ordered according to sections of the cone whose summit is occupied by the eye). Of course, a well-defined observer extracts everything that it can, everything that can be extracted in the corresponding system. In short, the role of a partial observer is *to perceive* and *to experience,* although these perceptions and affections are not those of a man, in the currently accepted sense, but belong to the things studied. Man feels the effect of them nonetheless (what mathematician does not fully experience the effect of a section, an ablation, or an addition), but he obtains this effect only from the ideal observer that he himself has installed like a golem in the system of reference. These partial observers belong to the neighborhood of the singularities of a curve, of a physical system, of a living organism. Even animism, when it multiplies little immanent souls in organs and functions, is not so far removed from biological science as it is said to be, on condition that these immanent souls are withdrawn from any active or efficient role so as to become solely sources of molecular perception and affection. In this way, bodies are populated by an infinity of little monads. The region of a state of affairs or a body apprehended by a partial observer will be called a site. Partial observers are forces. Force, however, is not what acts but, as Leibniz and Nietzsche knew, what perceives and experiences.

Wherever purely functional properties of recognition or selection appear, without direct action, there are observers: hence this is so throughout molecular biology, in immunology, or with allosteric enzymes.[12] Maxwell already presupposed a demon capable of distinguishing between rapid and slow molecules, between those with high and weak energy, within a mixture. It is true that in a system in a state of equilibrium, this demon of Maxwell's linked to the gas will necessarily be affected by vertigo; nonetheless it can spend a long time in a metastable state close to an enzyme. Particle physics needs countless infinitely subtle observers. We can conceive of partial observers whose site is smaller the more the state of affairs undergoes changes of coordinates. Finally, *ideal partial observers are the perceptions or sensory affections of functives themselves.* Even geometrical figures have affections and perceptions (pathemes and symptoms, said Proclus) without which the simplest problems would remain unintelligible. Partial observers are *sensibilia* that are doubles of the functives. Rather than oppose sensory knowledge and scientific knowledge, we should identify the sensibilia that populate systems of coordinates and are peculiar to science. This is what Russell did when he evoked those qualities devoid of all subjectivity, sense data distinct from all sensation, sites established in states of affairs, empty perspectives belonging to things themselves, contracted bits of space-time that correspond to the whole or to parts of a function. He assimilated them to apparatus and instruments like Michelson's interferometer or, more simply, the photographic plate, camera, or mirror that captures what no one is there to see and make these unsensed sensibilia blaze.[13] Far from these sensibilia being defined by instruments, since the latter are waiting for a real observer to come and see, it is instruments that presuppose the ideal partial observer situated at a good vantage point in things: the nonsubjective observer is precisely the sensory that qualifies (sometimes in a thousand ways) a scientifically determined state of affairs, thing, or body.

For their part, conceptual personae are philosophical sensibilia, the perceptions and affections of fragmentary concepts themselves: through them concepts are not only thought but perceived and felt. However, it is not enough to say that they are distinguished from scientific observers in the same way that concepts are distinguished from functives, since they would then contribute no further determination: both agents of enunciation must be distinguished not only by the perceived but by the mode of perception (nonnatural in both cases). It is not enough to assimilate the scientific observer (for example, the cannonball traveler of relativity) to a simple *symbol* that would mark states of variables, as Bergson does, while the philosophical persona would have the privilege of *the lived* (a being that endures) because he will undergo the variations themselves.[14] The philosophical persona is no more lived experience than the scientific observer is symbolic. There is ideal perception and affection in both, but they are very different from each other. Conceptual personae are always already on the horizon and function on the basis of infinite speed, nonenergetic differences between the rapid and the slow coming only from the surfaces they survey or from the components through which they pass in a single instant. Thus, perception does not transmit any information here, but circumscribes a (sympathetic or antipathetic) affect. Scientific observers, on the other hand, are points of view in things themselves that presuppose a calibration of horizons and a succession of framings on the basis of slowing-downs and accelerations: affects here become energetic relationships, and perception itself becomes a quantity of information. We cannot really develop these determinations because the status of pure percepts and affects, referring to the existence of the arts, has not yet been grasped. But, the fact that there are specifically philosophical perceptions and affections and specifically scientific ones—in short, sensibilia of the concept and sensibilia of the function—already indicates the basis of a relationship between science and philosophy, science and art, and philosophy and art, such that we can say that a function is beautiful and a concept is beautiful. The special perceptions and affections of science or philosophy necessarily connect up with the percepts and affects of art, those of science just as much as those of philosophy.

As for the direct confrontation of science and philosophy, it develops under three principle heads of opposition that group the series of functives on the one hand and the properties of concepts on the other. First there is the system of reference and plane of immanence; then independent variables and inseparable variations; and finally partial observers and conceptual personae. These are two types of multiplicity. A function can be given without the concept itself being given, although it can and must be; a function of space can be given without the concept of this space yet being given. The function in science determines a state of affairs, thing, or body that actualizes the virtual on a plane of reference and in a system of coordinates; the concept in philosophy expresses an event that gives consistency to the virtual on a plane of immanence and in an ordered form. In each case the respective fields of creation find themselves marked out by very different entities but that nonetheless exhibit a certain analogy in their tasks: a *problem,* in science or in philosophy, does not consist in answering a question but in adapting, in co-adapting, with a higher "taste" as problematic faculty, corresponding elements in the process of being determined (for example, for science, choosing the good independent variables, installing the effective partial observer on a particular route, and constructing the best coordinates of an equation or function). This analogy imposes two more tasks. How are we to conceive of practical transitions between the two sorts of problems? But above all, theoretically, do the heads of opposition rule out any uniformization and even any reduction of concepts to functives, or the other way around? And if no reduction is possible, how can we think a set of positive relations between the two?

Logic is reductionist not accidentally but essentially and necessarily: following the route marked out by Frege and Russell, it wants to turn the concept into a function. But this means first of all not only that the function must be defined in a mathematical or scientific proposition but that it characterizes a more general order of the proposition as what is expressed by the sentences of a natural language. Thus a new, specifically logical type of function must be invented. The propositional function "x is human" clearly shows the position of an independent variable that does not belong to the function as such but without which the function is incomplete. The complete function is made up of one or more "ordered pairs." A relation of dependence or correspondence (necessary reason) defines the function, so that "being human" is not itself the function, but the value of f(a) for a variable x. It hardly matters that most propositions have several independent variables or even that the notion of variable, insofar as it is linked to an indeterminate number, is replaced by that of argument, implying a disjunctive assumption within limits or an interval. The relation of the propositional function to the independent variable or argument defines the proposition's *reference,* or the function's truth value ("true" or "false") for the argument: John is a man, but Bill is a cat. The set of a function's truth values that determine true affirmative propositions constitutes a concept's *extension:* the concept's objects occupy the place of variables or arguments of the propositional function for which the proposition is true, or its reference satisfied. Thus the concept itself is the function for the set of objects that constitute its extension. In this sense every complete concept is a set and has a determinate number; the concept's objects are the *elements* of the set.[1]

It is still necessary to determine the conditions of reference that provide the limits or intervals into which a variable enters in a true proposition: x is a man, John is a man, because he did this, because he appears in this way. Such conditions of reference constitute not the concept's comprehension but its intension. They are presentations or logical descriptions, intervals, potentials, or "possible worlds," as the logicians say, coordinate axes, states of affairs or situations, the concept's *subsets:* evening star and morning star. For example, a concept with a single element, the concept of Napoleon I, has for its intension "the victor at Jena," "the one who was defeated at Waterloo." There is no qualitative difference between intension and extension here since both concern reference, intension being simply the condition of reference and constituting an endoreference of the proposition, extension constituting the exoreference. Reference is not left behind by ascending to its condition; we remain within extensionality. The question is rather one of knowing how, through these intensional presentations, we arrive at a univocal determination of objects or elements of the concept, of propositional variables, and of arguments of the function from the point of view of exoreference (or of the representation). This is the problem of proper names, and the business of a logical identification or individuation that takes us from states of affairs to the thing or body (object), through operations of quantification that also make possible attribution of the thing's essential predicates as that which finally constitutes the concept's *comprehension*. Venus (the evening star and morning star) is a planet that takes less time than the earth to complete its revolution. "Victor at Jena" is a description or presentation, whereas "general" is a predicate of Bonaparte, "emperor" a predicate of Napoleon, although being named general or holy emperor may be descriptions. The "propositional concept" therefore evolves entirely within the circle of reference insofar as it carries out a logicization of functives that thus become the prospects of a proposition (passage from the scientific to the logical proposition).

Sentences have no self-reference, as the paradox "I lie" shows. Not even performatives are self-referential but rather imply an exoreference of the proposition (the action that is linked to it by convention and accomplished by stating the proposition), and an endoreference (the status or state of affairs that entitles one to formulate the statement: for example, the concept's intension in the statement "I swear it" may be a witness in court, a child blamed for something, a lover declaring himself, etc.).[2] On the other hand, if we ascribe self-consistency to the sentence, this can only reside in the formal noncontradiction of the proposition or between propositions. But this means that propositions do not materially enjoy any endoconsistency or exoconsistency. To the extent that a cardinal number belongs to the propositional concept, the logic of propositions needs a scientific demonstration of the consistency of the arithmetic of whole numbers, on the basis of axioms. Now, according to the two aspects of Gödel's theorem, proof of the consistency of arithmetic cannot be represented within the system (there is no endoconsistency), and the system necessarily comes up against true statements that are nevertheless not demonstrable, are undecidable (there is no exoconsistency, or the consistent system cannot be complete). In short, *in becoming propositional, the concept loses all the characteristics it possessed as philosophi-*

cal concept: its self-reference, its endoconsistency and its exoconsistency. This is because a regime of independence (of variables, axioms, and undecidable propositions) has replaced that of inseparability. Even possible worlds as conditions of reference are cut off from the concept of the Other person that would give them consistency (so that logic finds itself oddly disarmed before solipsism). The concept in general no longer has a combination but an arithmetical number; the undecidable no longer indicates the inseparability of intensional components (zone of indiscernibility) but, on the contrary, the necessity of distinguishing them according to the requirement of reference, which renders all consistency (self-consistency) "uncertain." Number itself indicates a general principle of separation: "the concept 'letter of the word *Zahl*' separates *Z* from *a*, *a* from *h*, etc." Functions derive all their power from reference, whether this be reference to states of affairs, things, or other propositions: reduction of the concept to the function inevitably deprives it of all its specific characteristics that referred back to another dimension.

Acts of reference are finite movements of thought by which science constitutes or modifies states of affairs and bodies. Historical man may also be said to carry out such modifications, but under conditions of the lived, where functives are replaced by perceptions, affections, and actions. The position is no longer the same with logic: since it considers empty reference in itself as simple truth value, it can only apply it to already constituted states of affairs or bodies, in established scientific propositions or in factual propositions (Napoleon is the one who was defeated at Waterloo) or in simple opinions ("*X* thinks that . . ."). All types of propositions are *prospects,* with an information value. Logic has therefore a paradigm, it is even the third case of paradigm, which is no longer that of religion or science but like *the recognition of truth* in prospects or informative propositions. The technical expression "metamathematics" clearly shows the passage from scientific statement to logical proposition in a form of recognition. The projection of this paradigm means that logical concepts are in turn only figures and that logic is an ideography. The logic of propositions needs a method of projection, and Gödel's theorem itself invents a projective model.[3] It is like an ordered, oblique deformation of reference in relation to its scientific status. Logic seems to be forever struggling with the complex question of how it differs from psychology. However, we can definitely agree that it sets up as a model an image by right of thought that is in no way psychological (without, for all that, being normative). The question lies rather in the value of this image by right and in what it claims to teach us about the mechanisms of a pure thought.

Of all the finite movements of thought, the form of recognition is certainly the one that goes the least far and is the most impoverished and puerile. From earliest times philosophy has encountered the danger of evaluating thought by reference to such uninteresting cases as saying "hello, Theodore" when Theatetus is passing by. The classical image of thought was not safe from these endeavors that value recognition of truth. It is hard to believe that the problems of thought, in science as well as in philosophy, are troubled by such cases: as the creation of thought, a problem has nothing to do with a question, which is only a suspended proposition, the bloodless double of an affirmative proposition that is supposed to serve as its answer ("Who is the author of *Waverley?*" "Is Scott the author of *Waverley?*"). Logic is always defeated by itself, that is to say, by the insignificance of the cases on which it thrives. In its desire to supplant philosophy, logic detaches the proposition from all its psychological dimensions, but clings all the more to the set of postulates that limited and subjected thought to the constraints of a recognition of truth in the proposition.[4] And when logic ventures into a calculus of problems, it does so by modeling it, isomorphically, on the calculus of propositions. It is less like a game of chess, or a language game, than a television quiz game. But problems are never propositional.

Instead of a string of linked propositions, it would be better to isolate the flow of interior monologue, or the strange forkings of the most ordinary conversation. By separating them from their psycho-logical, as well as their sociological adhesions, we would be able to show how thought as such produces something *interesting* when it accedes to the infinite movement that frees it from truth as supposed paradigm and reconquers an immanent power of creation. But to do this it would be necessary to return to the interior of scientific states of affairs or bodies in the process of being constituted, in order to penetrate into consistency, that is to say, into the sphere of the virtual, a sphere that is only actualized in them. *It would be necessary to go back up the path that science descends,* and at the very end of which logic sets up its camp (the same goes for History, where we would have to arrive at the unhistorical vapor that goes beyond actual factors to the advantage of a creation of something new). But it is this sphere of the virtual, this Thought-nature, that logic can only *show,* according to a famous phrase, without ever being able to grasp it in propositions or relate it to a reference. Then logic is silent, and it is only interesting when it is silent. Paradigm for paradigm, it is then in agreement with a kind of Zen Buddhism.

By confusing concepts with functions, logic acts as though science were already dealing with concepts or forming concepts of the first zone. But it must itself double scientific with logical functions that are supposed to form a new class of purely logical, or second zone, concepts. A real hatred inspires logic's rivalry with, or its will to supplant, philosophy. It kills the concept twice over. However, the concept is reborn because it is not a scientific function and because it is not a logical proposition: it does not belong to a discursive system and it does not have a reference. The concept shows itself and does nothing but show itself. Concepts are really monsters that are reborn from their fragments.

Logic itself sometimes allows philosophical concepts to reappear, but in what form and state? As concepts in general have found a pseudorigorous status in scientific and logical functions, philosophy inherits *concepts of the third zone* that are outside number and no longer constitute clearly demarcated and well-defined sets that can be related to mixtures ascribable as physico-mathematical states of affairs. They are, instead, vague or fuzzy sets, simple aggregates of perceptions and affections, which form within the lived as immanent to a subject, to a consciousness. They are qualitative or intensive multiplicities, like "redness" or "baldness," where we cannot decide whether certain elements do or do not belong to the set. These lived sets are expressed in a third kind of prospects, which are no longer those of scientific statements or logical propositions but of the subject's pure and simple opinions, of subjective evaluations or judgments of taste: this is already red, he is nearly bald. However, even for an enemy of philosophy, the refuge of philosophical concepts cannot immediately be found in such empirical judgments. We must isolate the functions, of which these fuzzy sets, these lived contents, are only variables. And at this point we face an alternative: *either* we will end up reconstituting scientific or logical functions for these variables, which would make the appeal to philosophical concepts definitively useless,[5] *or* we will have to invent a new, specifically philosophical type of function, a third zone in which everything seems to be strangely reversed, since it will be given the task of supporting the other two.

If the world of the lived is like the earth, which must found and support the science and logic of states of affairs, it is clear that apparently philosophical concepts are required to carry out this first foundation. The philosophical concept thus requires a "belonging" to a subject and no longer to a set. Not that the philosophical concept is to be confused with the merely lived, even if it is defined as a multiplicity of fusion or as immanence of a flow to the subject—the lived only furnishes variables, whereas concepts must still define true functions. These functions will have reference only to the lived, as scientific functions have reference to states of affairs. Philosophical concepts will be functions of the lived, as scientific concepts are functions of states of affairs; but the order or the derivation now changes direction since these functions of the lived become primary. A transcendental logic (it can also be called dialectical) embraces the earth and all that it bears, and this serves as the primordial ground for formal logic and the derivative regional sciences. It is necessary therefore to discover at the very heart of the immanence of the lived to a subject, that subject's acts of transcendence *capable of constituting new functions of variables or conceptual references:* in this sense the subject is no longer solipsist and empirical but transcendental. We have seen that Kant began to accomplish this task by showing how philosophical concepts are necessarily related to lived experience through a priori propositions or judgment as functions of a whole of possible experience. But it is Husserl who sees it through to the end by discovering, in non-numerical multiplicities or immanent per- ceptivo-affective fusional sets, the triple root of acts of transcendence (thought) through which the subject constitutes first of all a sensory world filled with objects, then an intersubjective world occupied by the other, and finally a common ideal world that will be occupied by scientific, mathematical, and logical formations. Numerous phenome- nological or philosophical concepts (such as "being in the world," "flesh," "ideality," etc.) are the expression of these acts. They are not only liveds that are immanent to the solipsist subject but references of the transcendental subject to the lived; they are not perceptivo- affective variables, but major functions which find in these variables their respective trajectories of truth. They are not vague or fuzzy sets, subsets, but totalizations that exceed all power of sets. They are not merely empirical judgments or opinions but proto-beliefs, *Urdoxa, original opinions as propositions.*[6] They are not successive contents of the flow of immanence but acts of transcendence that traverse it and carry it away by determining the "significations" of the potential totality of the lived. The concept as signification is all of this at once: immanence of the lived to the subject, act of transcendence of the subject in relation to variations of the lived, totalization of the lived or function of these acts. It is as if philosophical concepts get going only by accepting to become special functions and by denaturing the immanence that they still need: as immanence is now only that of the lived it is inevitably immanence to a subject, whose acts (functions) will be concepts relative to this lived—following, as we have seen, the long denaturation of the plane of immanence.

Although it may be dangerous for philosophy to depend on the generosity of logicians, or on their regrets, we might wonder whether a precarious balance cannot be found between scientifico-logical concepts and phenomenological-philosophical concepts. Gilles-Gaston Granger has suggested a distribution in which the concept, being determined first of all as a scientific or logical concept, nonetheless allows a place for philosophical functions in a third but autonomous zone, for functions or significations of the lived as virtual totality (fuzzy sets seem to play the role of a hinge between the two forms of concepts).[7] Science has therefore arrogated the concept to itself, but there are nevertheless nonscientific, that is to say, phenomenological concepts that are tolerated in homeopathic doses—hence the strangest hybrids of Frego-Husserlianism, or even Wittgensteino-Heideggerianism, that we see springing up today. This has long been the situation of philosophy in America, with a large department of logic and a very small one of phenomenology, even though the two parts were usually at war. It is like the proverbial lark pie containing one lark and one horse. But the phenomenological lark is not even the most exquisite portion; it is only what the logical horse sometimes leaves for philosophy. The situation is more like the rhinoceros and the bird that lives on its parasites.

There is a long series of misunderstandings about the concept. It is true that the concept is fuzzy or vague not because it lacks an outline but because it is vagabond, nondiscursive, moving about on a plane of immanence. It is intensional or modular not because it has conditions of reference but because it is made up of inseparable variations that pass through zones of indiscernibility and change its outline. It has no reference at all, either to the lived or to states of affairs, but a consistency defined by its internal components. The concept is neither denotation of states of affairs nor signification of the lived; it is the event as pure sense that immediately runs through the components. It has no number, either whole or fractional, for counting things that display its properties, but a combination that condenses and accumulates the components it traverses and surveys. The concept is a form or a force; in no possible sense is it ever a function. In short, there are only philosophical concepts on the plane of immanence, and scientific functions or logical propositions are not concepts.

Prospects designate first of all the elements of the proposition (propositional function, variables, truth value), but also the various types of propositions or modalities of judgment. If the philosophical concept is confused with a function or a proposition, it is not as a scientific or even logical kind but, by analogy, as a function of the lived or a proposition of opinion (third type). Hence a concept must be produced that takes account of this situation: what *opinion* proposes is a particular relationship between an external perception as state of a subject and an internal affection as passage from one state to another (exo- and endoreference). We pick out a quality supposedly common to several objects that we perceive, and an affection supposedly common to several subjects who experience it and who, along with us, grasp that quality. Opinion is the rule of the correspondence of one to the other; *it is a function or a proposition whose arguments are perceptions and affections,* and in this sense it is a function of the lived. For example, we grasp a perceptual quality common to cats or dogs and a certain feeling that makes us like or hate one or the other: for a group of objects we can extract many diverse qualities and form many groups of quite different, attractive or repulsive, subjects (the "society" of those who like cats or detest them), so that opinions are essentially the object of a struggle or an exchange. This is the Western democratic, popular conception of philosophy as providing pleasant or aggressive dinner conversations at Mr. Rorty's. Rival opinions at the dinner table—is this not the eternal Athens, our way of being Greek again? The three characteristics by which philosophy was related to the Greek city were, precisely, the society of friends, the table of immanence, and the confrontation of opinions. One might object that Greek philosophers were always attacking *doxa* and contrasting it with an episteme as the only knowledge adequate to philosophy. But this is a mixed-up business, and philosophers, being only friends and not wise men, find it very difficult to give up *doxa*.

Doxa is a type of proposition that arises in the following way: in a given perceptive-affective lived situation (for example, some cheese is brought to the dinner table), someone extracts a pure quality from it (for example, a foul smell); but, at the same time as he abstracts the quality, he identifies himself with a generic subject experiencing a common affection (the society of those who detest cheese—competing as such with those who love it, usually on the basis of another quality). "Discussion," therefore, bears on the choice of the abstract perceptual quality and on the power of the generic subject affected. For example, is to detest cheese to manage without being a bon vivant? But is being a bon vivant a generically enviable affection? Ought we not say that it is those who love cheese, and all bons vivants, who stink? Unless it is the enemies of cheese who stink. This is like the story, told by Hegel, of the shopkeeper to whom it was said, "Your eggs are rotten old woman," and who replied, "Rot yourself, and your mother, and your grandmother": opinion is an abstract thought, and insult plays an effective role in this abstraction because opinion expresses the general functions of particular states.[8] It extracts an abstract quality from perception and a general power from affection: in this sense all opinion is already political. That is why so many discussions can be expressed in this way: "as a man, I consider all women to be unfaithful"; "as a woman, I think men are liars."

Opinion is a thought that is closely molded on the form of recognition—recognition of a quality in perception (contemplation), recognition of a group in affection (reflection), and recognition of a rival in the possibility of other groups and other qualities (communication). It gives to the recognition of truth an extension and criteria that are naturally those of an "orthodoxy": a true opinion will be the one that coincides with that of the group to which one belongs by expressing it. This is clear to see in certain competitions: you must express your opinion, but you "win" (you have spoken the truth) if you say the same as the majority of those participating in the competition. The essence of opinion is will to majority and already speaks in the name of a majority. Even the man of "paradoxes" only expresses himself with so many winks and such stupid self-assurance because he claims to express everyone's secret opinion and to be the spokesman of that which others dare not say. This is still only the first step of opinion's reign: opinion triumphs when the quality chosen ceases to be the condition of a group's constitution but is now only the image or "badge" of the constituted group that itself determines the perceptive and affective model, the quality and affection, that each must acquire. Then marketing appears as the concept itself: "We, the conceivers. . . ." Ours is the age of communication, but every noble soul flees and crawls far away whenever a little discussion, a colloquium, or a simple conversation is suggested. In every conversation the fate of philosophy is always at stake, and many philosophical discussions do not as such go beyond discussions of cheese, including the insults and the confrontation of worldviews. The philosophy of communication is exhausted in the search for a universal liberal opinion as consensus, in which we find again the cynical perceptions and affections of the capitalist himself.

EXAMPLE I I

How does this situation concern the Greeks? It is often said that since Plato, the Greeks contrasted philosophy, as a *knowledge* that also includes the sciences, with *opinion-doxa,* which they relegate to the sophists and rhetors. But we have learned that the opposition was not so clear-cut. How could philosophers possess knowledge, philosophers who cannot and do not want to restore the knowledge of the sages and who are only friends? And how, since it takes a truth value, could opinion be something entirely for the sophists?[9]

Furthermore, it seems that the Greeks had a clear-enough idea of science, which was not confused with philosophy: it was a knowledge of the cause, of the definition, a sort of function already. So, the whole problem was, How can one arrive at definitions, at these premises of the scientific or logical syllogism? It was by means of the dialectic: an investigation that aimed, on a given theme, to determine which opinions were the most plausible by reference to the quality they extracted and which opinions were the wisest by reference to the subject who advanced them. Even in Aristotle, the dialectic of opinions was necessary for determining possible scientific propositions, and in Plato "true opinion" was a prerequisite for knowledge and the sciences. Even Parmenides did not pose knowledge and opinion as being two disjunctive pathways.[10] Whether or not they were democrats, the Greeks did not so much oppose knowledge and opinion as fight over opinions, as confront and compete against each other in the element of pure opinion. Philosophers blamed the sophists not for confining themselves to *doxa* but for making a bad choice of the quality to be extracted from perceptions, and of the generic subject to be isolated from affections, so that the sophists could not reach what was "true" in an opinion: they remained prisoners of variations of the lived. Philosophers blamed the sophists for being content with any kind of sensory quality in relation to an individual man, or to mankind, or to the nomos of the city (three interpretations of Man as power or the "measure of everything"). But Platonist philosophers themselves had an extraordinary answer that, they thought, allowed them to select opinions. *It was necessary to choose the quality that was like the unfolding of the Beautiful in any lived situation,* and to take as generic subject Man inspired by the Good. Things had to unfold in the beautiful, and their users be inspired by the good, for opinion to achieve the Truth. This was not always easy. The beautiful in Nature and the good in minds define philosophy as a function of the variable life. Thus Greek philosophy is the moment of the beautiful; the beautiful and the good are functions whose truth value is opinion. To reach true opinion, perception had to be taken as far as the beauty of the perceived (*dokounta*) and affection as far as the test of the good (*dokimôs*): this will no longer be changing and arbitrary opinion but *an original opinion, a proto-opinion* that restores to us the forgotten homeland of the concept as, in the great Platonic trilogy, the love of the *Symposium,* the delirium of the *Phaedrus,* and the death of the *Phaedo.* Where, on the contrary, the sensory appears without beauty, reduced to illusion, and the mind appears without good, given over to simple pleasure, opinion will remain sophistical itself and false—cheese, perhaps, or mud or hair. However, does not this passionate search for true opinion lead the Platonists to an aporia, the very one expressed in the most astonishing dialogue, the *Theatetus?* Knowledge must be transcendent; it must be added to and distinguished from opinion in order to make opinion true. But knowledge must be immanent for opinion to be true as opinion. Greek philosophy still remains attached to that old Wisdom ready to unfold its transcendence again, although it now possesses only its friendship, its affection. Immanence is necessary, but it must be immanent to something transcendent, to ideality. The beautiful and the good continue to lead us back to transcendence. It is as if true opinion still demanded a knowledge that it had nevertheless deposed.

Phenomenology can be seen as taking up a similar task. It, too, goes in search of original opinions which bind us to the world as to our homeland (earth). It also needs the beautiful and the good so that the latter are not confused with variable empirical opinion and so that perception and affection attain their truth value. This time it is a question of the beautiful in art and the constitution of humanity in history. Phenomenology needs art as logic needs science; Erwin Straus, Merleau-Ponty, or Maldiney need Cézanne or Chinese painting. The lived turns the concept into nothing more than an empirical opinion as psychosociological type. The immanence of the lived to a transcendental subject, therefore, must turn opinion into a proto-opinion in whose constitution art and culture are involved and that is expressed as an act of transcendence of this subject within the lived (communication), so as to form a community of friends. But does not the Husserlian transcendental subject hide European man whose privilege it is constantly to "Europeanize," as the Greeks "Greekized," that is to say, to go beyond the limits of other cultures that are preserved as psychosocial types? Are we not led back in this way to the simple opinion of the average Capitalist, the great Major, the modern Ulysses whose perceptions are clichés and whose affections are labels, in a world of communication that has become marketing and from which not even Cézanne or Van Gogh can escape? The distinction between original and derivative is not by itself enough to get us out of the simple domain of opinion, and the *Urdoxa* does not raise us to the level of the concept. As in the Platonic aporia, phenomenology is never more in need of a higher wisdom, of a "rigorous science," than when it invites us to renounce it. Phenomenology wanted to renew our concepts by giving us perceptions and affections that would awaken us to the world, not as babies or hominids but as, by right, beings whose proto-opinions would be the foundations of this world. But we do not fight against perceptual and affective clichés if we do not also fight against the machine that produces them. By invoking the primordial lived, by making immanence an immanence to a subject, phenomenology could not prevent the subject from forming no more than opinions that already extracted clichés from new perceptions and promised affections. We will continue to evolve in the form of recognition; we will invoke art, but without reaching the concepts capable of confronting the artistic affect and percept. The Greeks with their cities, and phenomenology with our Western societies, are certainly right to consider opinion as one of the conditions of philosophy. But, by invoking art as the means of deepening opinion and of discovering original opinions, will philosophy find the path that leads to the concept? Or should we, along with art, overturn opinion, raising it to the infinite movement that replaces it with, precisely, the concept?

Confusing the concept with the function is ruinous for the philosophical concept in several respects. It makes science the concept par excellence, which is expressed in the scientific proposition (first prospect). It replaces the philosophical concept with a logical concept, which is expressed in factual propositions (second prospect). It leaves the philosophical concept with a reduced or defective share that it carves out in the domain of opinion (third prospect) by exploiting its friendship with a higher wisdom or with a rigorous science. But the concept's place is not in any of these three discursive systems. The concept is no more a function of the lived than it is a scientific or logical function. We discover the irreducibility of concepts to functions only if, instead of setting them against one another in an indeterminate way, we compare what constitutes the reference of one with what produces the other's consistency. *States of affairs, objects* or *bodies,* and *lived states* form the function's references, whereas *events* are the concept's consistency. These are the terms that have to be considered from the point of view of a possible reduction.

EXAMPLE 12

In contemporary thought such a comparison seems to correspond to Badiou's particularly interesting undertaking. He proposes to distribute at intervals on an ascending line a series of factors passing from functions to concepts. He takes a base neutralized in relation to both concepts and functions—any multiplicity whatever that is presented as a Set that can be raised to infinity. The first instance is the *situation,* when the set is related to elements that are doubtless multiplicities but that are subject to a regime of the "counting as one" (bodies or objects, units of the situation). In the second place, *situation-states* are subsets, always exceeding elements of the set or objects of the situation; but this excess of the state no longer lets itself be hierarchized as in Cantor—it is "inassignable," following an "errant line," in conformity with the development of set theory. It must still be re-presented in the situation, this time as "indiscernible" at the same time that the situation becomes almost complete: the errant line here forms four figures, four loops as *generic functions*—scientific, artistic, political or doxic, and amorous or lived—to which productions of "truths" correspond. But perhaps we then arrive at a conversion of immanence of the situation, a conversion of the excess to the void, which will reintroduce the transcendent: this is the *event site* that sticks to the edge of the void in the situation and now includes not units but singularities as elements dependent on the preceding functions. Finally, the *event* itself appears (or disappears), less as a singularity than as a separated aleatory point that is added to or subtracted from the site, within the transcendence of the void or *the* truth as void, without it being possible to decide on the adherence of the event to the situation in which it finds its site (the undecidable). On the other hand, perhaps there is an operation like a dice throw on the site that qualifies the event and makes it enter into the situation, a power of "making" the event. The fact that the event is the concept, or philosophy as concept, distinguishes it from the four preceding functions, although it takes conditions from them and imposes conditions on them in turn—that art is fundamentally "poem," that science is set-theoretical [*ensembliste*], that love is the unconscious of Lacan, and that politics escapes from opinion-*doxa*.[11]

By starting from a neutralized base, the set, which indicates any multiplicity whatever, Badiou draws up a line that is single, although it may be very complex, on which functions and concepts will be spaced out, the latter above the former: philosophy thus seems to float in an empty transcendence, as the unconditioned concept that finds the totality of its generic conditions in the functions (science, poetry, politics, and love). Is this not the return, in the guise of the multiple, to an old conception of the higher philosophy? It seems to us that the theory of multiplicities does not support the hypothesis of any multiplicity whatever (even mathematics has had enough of set-theoreticism [*ensemblisme*]). There must be at least two multiplici*ties,* two types, from the outset. This is not because dualism is better than unity but because the multiplicity is precisely what happens between the two. Hence, the two types will certainly not be one above the other but rather one beside the other, against the other, face to face, or back to back. Functions and concepts, actual states of affairs and virtual events, are two types of multiplicities that are not distributed on an errant line but related to two vectors that intersect, one according to which states of affairs actualize events and the other according to which events absorb (or rather, adsorb) states of affairs.

States of affairs leave the virtual chaos on conditions constituted by the limit (reference): they are actualities, even though they may not yet be bodies or even things, units, or sets. They are masses of independent variables, particles-trajectories or signs-speeds. They are *mixtures*. These variables determine singularities, insofar as they enter into coordinates, and are held within relations according to which one of them depends upon a large number of others or, conversely, many of them depend upon one. A potential or power is found to be associated with such a state of affairs (the importance of the Leibnizian formula $mv2$ is due to its introducing a potential into the state of affairs). This is because the state of affairs actualizes a chaotic virtuality by carrying along with it a space that has ceased, no doubt, to be virtual but that still shows its origin and serves as absolutely indispensable correlate to the state of affairs. For example, in the actuality of the atomic nucleus, the nucleon is still close to chaos and finds itself surrounded by a cloud of constantly emitted and reabsorbed particles; but at a further level of actualization, the electron is in relation with a potential photon that interacts with the nucleon to give a new state of the nuclear material. *A state of affairs cannot be separated from the potential through which it takes effect* and without which it would have no activity or development (for example, catalysis). It is through this potential that it can confront accidents, adjunctions, ablations, or even projections, as we see in geometrical figures: either losing and gaining variables, extending singularities up to the neighborhood of new ones, or following bifurcations that transform it, or passing through a phase space whose number of dimensions increases with supplementary variables, or, above all, individuating bodies in the field that it forms with the potential. None of these operations come about all by themselves; they all constitute "problems." It is the privilege of the living being to reproduce from within the associated potential in which it actualizes its state and individualizes its body. But an essential moment in every domain is the passage from a state of affairs to the body through the intermediary of a potential or power or, rather, the division of individuated bodies within the subsisting state of affairs. We pass here from mixture to *interaction*. And finally, the interactions of bodies condition a sensibility, a proto-perceptibility and a proto-affectivity that are already expressed in the partial observers attached to the state of affairs, although they complete their actualization only in the living being. What is called "perception" is no longer a state of affairs but a state of the body as induced by another body, and "affection" is the passage of this state to another state as increase or decrease of potential-power through the action of other bodies. Nothing is passive, but everything is interaction, even gravity. This was the definition Spinoza gave of "affectio" and "affectus" for bodies grasped within a state of affairs, and that Whitehead rediscovered when he made each thing a "prehension" of other things and the passage from one prehension to another a positive or negative "feeling."* Interaction becomes *communication*. The ("public") matter of fact was the mixture of data actualized by the world in its previous state, while bodies are new actualizations whose "private" states restore matters of fact for new bodies.[12] Even when they are nonliving, or rather inorganic, things have a lived experience because they are perceptions and affections.

When philosophy compares itself to science, it sometimes puts forward a simplistic image of the latter, which makes scientists laugh. However, even if philosophy has the right to offer an image of science (through concepts) that has no scientific value, it has nothing to gain

---

*In English in the original.

by attributing limits to science that scientists continually go beyond in their most elementary procedures. Thus, when philosophy relegates science to the "already-made" and reserves for itself the "being-made," like Bergson or phenomenology, and particularly in Erwin Straus, we not only run the risk of assimilating philosophy to a simple lived but give a bad caricature of science: Paul Klee's vision was certainly more sound when he said that mathematics and physics, in addressing themselves to the functional, take not the completed form but formation itself as their object.[13] Furthermore, when we compare philosophical and scientific multiplicities, conceptual and functional multiplicities, it may be much too simple to define the latter by sets. Sets, as we have seen, are of interest only as actualization of the limit; they depend on functions and not the converse, and the function is the true object of science.

Functions are, first of all, functions of states of affairs and thus constitute scientific propositions as the first type of prospects: their arguments are independent variables on which coordinations and potentializations are carried out that determine their necessary relations. In the second place, functions are functions of things, objects, or individuated bodies that constitute logical propositions: their arguments are singular terms taken as independent logical atoms on which descriptions are brought to bear (logical states of affairs) that determine their predicates. In the third place, the arguments of functions of the lived are perceptions and affections, and they constitute opinions (*doxa* as third type of prospect): we have opinions on everything that we see or that affects us, to the extent that the human sciences can be seen as a vast doxology—but things themselves are generic opinions insofar as they have molecular perceptions and affections, in the sense that the most elementary organism forms a proto-opinion on water, carbon, and salts on which its conditions and power depend. Such is the path that descends from the virtual to states of affairs and to other actualities: we encounter no concepts on this path, only functions. *Science passes from chaotic virtuality to the states of affairs and bodies that actualize it.* However, it is inspired less by the concern for unification in an ordered actual system than by a desire not to distance itself too much from chaos, to seek out potentials in order to seize and carry off a part of that which haunts it, the secret of the chaos behind it, the pressure of the virtual.[14]

Now, if we go back up in the opposite direction, from states of affairs to the virtual, the line is not the same because it is not the same virtual (we can therefore go down it as well without it merging with the previous line). The virtual is no longer the chaotic virtual but rather virtuality that has become consistent, that has become an entity formed on a plane of immanence that sections the chaos. This is what we call the Event, or the part that eludes its own actualization in everything that happens. The event is not the state of affairs. It is actualized in a state of affairs, in a body, in a lived, but it has a shadowy and secret part that is continually subtracted from or added to its actualization: in contrast with the state of affairs, it neither begins nor ends but has gained or kept the infinite movement to which it gives consistency. It is the virtual that is distinct from the actual, but a virtual that is no longer chaotic, that has become consistent or real on the plane of immanence that wrests it from the chaos— it is a virtual that is real without being actual, ideal without being abstract. The event might seem to be transcendent because it surveys the state of affairs, but it is pure immanence that gives it the capacity to survey itself by itself and on the plane. What is transcendent, transdescendent, is the state of affairs in which the event is actualized. But, even in this state of affairs, the event is pure immanence of what is not actualized or of what remains indifferent to actualization, since its reality does not depend upon it. The event is immaterial, incorporeal, unlivable: pure *reserve*. Two thinkers have gone the farthest into the event—Péguy and Blanchot. Blanchot says that it is necessary to distinguish between, on the one hand, the accomplished or potentially accomplished state of affairs in an at least potential relation with my body, with myself; and, on the other hand, the event, that its own reality cannot bring to completion, the interminable that neither stops nor begins, that remains without relation to myself, and my body without relation to it—infinite movement. Péguy says that it is necessary to distinguish between, on the one hand, the state of affairs through which we, ourselves, and our bodies, pass and, on the other hand, the event into which we plunge or return, that which starts again without ever having begun or ended—the immanent aternal [*l'internel*].[15]

Throughout a state of affairs, a cloud or a flow, even, we seek to isolate variables at this or that instant, to see when, on the basis of a potential, new ones arise, into what relations of dependence they can enter, through what singularities they pass, what thresholds they cross, and what bifurcations they take. We mark out the functions of the state of affairs: differences between the local and the global are internal to the domain of functions (for example, depending on whether all independent variables but one can be eliminated). *The differences between the physico-mathematical, the logical, and the lived* also pertain to functions (depending on whether bodies are grasped in the singularities of states of affairs, or as themselves singular terms, or according to singular thresholds between perception and affection). An actual system, a state of affairs, or a domain of functions are at any rate defined as a time between two instants, or as times between many instants. That is why, when Bergson says that there is always time between two instants, however close to each other they may be, he has still not left the domain of functions and introduces only a little of the lived into it.

But when we ascend toward the virtual, when we turn ourselves toward the virtuality that is actualized in the state of affairs, we discover a completely different reality where we no longer have to search for what takes place from one point to another, from one instant to another, because virtuality goes beyond any possible function. In the conversational words attributable to a scientist, the event "doesn't care where it is, and moreover it doesn't care how long it's been going," so that art and even philosophy may apprehend it better than science.[16] It is no longer time that exists between two instants; it is the event that is a meanwhile [*un entre-temps**]: the meanwhile is not part of the eternal, but neither is it part of time—it belongs to becoming. The meanwhile, the event, is always a dead time; it is there where nothing takes place, an infinite awaiting that is already infinitely past, awaiting and reserve. This dead time does not come after what happens; it coexists with the instant or time of the accident, but as the immensity of the empty time in which we see it as still to come and as having already happened, in the strange indifference of an intellectual intuition. All the meanwhiles are superimposed on one another, whereas times succeed each other. In every event there are many heterogeneous, always simultaneous components, since each of them is a meanwhile, all within the meanwhile that makes them communicate through zones of indiscernibility, of undecidability: they are variations, modulations, intermezzi, singularities of a new infinite order. Each component of the event is *actualized or effectuated* in an instant, and the event in the time that passes between these instants; but nothing happens within *the virtuality* that has only meanwhiles as components and an event as composite becoming. Nothing happens there, but everything becomes, so that the event has the privilege of beginning again when time is past.[17] Nothing happens, and yet everything changes, because becoming continues to pass through its components again and to restore the event that is actualized elsewhere, at a different moment. When time passes and takes the instant away, there is always a meanwhile to restore the event. It is a *concept* that apprehends the event, its becoming, its inseparable variations; whereas a function grasps a state of affairs, a time and variables, with their relations depending on time. The

---

*We have followed the usual translation of *entre-temps* as signifying "meanwhile" or "meantime," although the English loses something of the literal meaning of the French as that which happens in the interval between moments of time or actions.

concept has a power of repetition that is distinct from the discursive power of the function. In its production and reproduction, the concept has the reality of a virtual, of an incorporeal, of an impassible, in contrast with functions of an actual state, body functions, and lived functions. Setting up a concept is not the same thing as marking out a function, although on both sides there is movement, and in each case there are transformations and creations: the two types of multiplicities intersect.

No doubt, the event is not only made up from inseparable variations, it is itself inseparable from the state of affairs, bodies, and lived reality in which it is actualized or brought about. But we can also say the converse: the state of affairs is no more separable from the event that nonetheless goes beyond its actualization in every respect. It is necessary to go back up to the event that gives its virtual consistency to the concept, just as it is necessary to come down to the actual state of affairs that provides the function with its references. From everything that a subject may live, from its own body, from other bodies and objects distinct from it, and from the state of affairs or physico-mathematical field that determines them, the event releases a vapor that does not resemble them and that takes the battlefield, the battle, and the wound as components or variations of a pure event in which there remains only an allusion to what concerns our states. The event is actualized or effectuated whenever it is inserted, willy-nilly, into a state of affairs; but it is *counter-effectuated* whenever it is abstracted from states of affairs so as to isolate its concept. There is a dignity of the event that has always been inseparable from philosophy as *amor fati:* being equal to the event, or becoming the offspring of one's own events—"my wound existed before me; I was born to embody it."[18] I was born to embody it as event because I was able to disembody it as state of affairs or lived situation. There is no other ethic than the *amor fati* of philosophy. Philosophy is always meanwhile. Mallarmé, who counter-effectuated the event, called it Mime because it side-steps the state of affairs and "confines itself to perpet-

ual allusion without breaking the ice."[19] Such a mime neither repro-
duces the state of affairs nor imitates the lived; it does not give an
image but constructs the concept. It does not look for the function of
what happens but extracts the event from it, or that part that does
not let itself be actualized, the reality of the concept. Not willing
what happens, with that false will that complains, defends itself and
loses itself in gesticulations, but taking the complaint and rage to the
point that they are turned against what happens so as to set up the
event, to isolate it, to extract it in the living concept. Philosophy's
sole aim is to become worthy of the event, and it is precisely the
conceptual persona who counter-effectuates the event. Mime is an
ambiguous name. It is he or she, the conceptual persona carrying out
the infinite movement. Willing war against past and future wars, the
pangs of death against all deaths, and the wound against all scars, in
the name of becoming and not of the eternal: it is only in this sense
that the concept gathers together.

From virtuals we descend to actual states of affairs, and from states
of affairs we ascend to virtuals, without being able to isolate one from
the other. But we do not ascend and descend in this way on the same
line: actualization and counter-effectuation are not two segments of
the same line but rather different lines. If we restrict ourselves to the
scientific functions of states of affairs, it seems that they cannot be
isolated from a virtual that they actualize, but this virtual appears first
of all as a cloud or a fog, or even as a chaos—as a chaotic virtuality
rather than the reality of an ordered event in the concept. That is
why it often appears to science that philosophy covers up a simple
chaos, leading science to say, "Your only choice is between chaos and
me, science." The line of actuality lays out a plane of reference that
slices the chaos again: it takes from it states of affairs that, of course,
also actualize virtual events in their coordinates, but it retains only
potentials already in the course of being actualized, forming part of
the functions. Conversely, if we consider philosophical concepts of
events, their virtuality refers to the chaos, but on a plane of imma-

nence that slices it again in turn, and that extracts from it only the consistency or reality of the virtual. No doubt states of affairs that are too dense are adsorbed, counter-effectuated by the event, but we find only allusions to them on the plane of immanence and in the event. The two lines are therefore inseparable but independent, each complete in itself: it is like the envelopes of the two very different planes. Philosophy can speak of science only by allusion, and science can speak of philosophy only as of a cloud. If the two lines are inseparable it is in their respective sufficiency, and philosophical concepts act no more in the constitution of scientific functions than do functions in the constitution of concepts. It is in their full maturity, and not in the process of their constitution, that concepts and functions necessarily intersect, each being created only by their specific means—a plane, elements, and agents in each case. That is why it is always unfortunate when scientists do philosophy without really philosophical means or when philosophers do science without real scientific means (we do not claim to have been doing this).

The concept does not reflect on the function any more than the function is applied to the concept. Concept and function must intersect, each according to its line. Riemannian functions of space, for example, tell us nothing about a Riemannian concept of space peculiar to philosophy: it is only to the extent that philosophy is able to create it that we have the concept of a function. In the same way, the irrational number is defined by a function as the common limit of two series of rational numbers, one of which has no maximum and the other no minimum. The concept, on the other hand, refers not to series of numbers but to strings of ideas that are reconnected over a lacuna (rather than linked together by continuation). Death may be assimilated to a scientifically determinable state of affairs, as a function of independent variables or even as one of the lived state, but it also appears as a pure event whose variations are coextensive with life: both, very different aspects are found in Bichat. Goethe constructs an imposing concept of color, with inseparable variations of light and shade, zones of indiscernibility, and processes of intensification that show the extent to which there is also experimentation in philosophy; whereas Newton constructed the function of independent variables or frequency. If philosophy has a fundamental need for the science that is contemporary with it, this is because science constantly intersects with the possibility of concepts and because concepts necessarily involve allusions to science that are neither examples nor applications, nor even reflections. Conversely, are there functions—properly scientific functions—of concepts? This amounts to asking whether science is, as we believe, equally and intensely in need of philosophy. But only scientists can answer that question.

The young man will smile on the canvas for as long as the canvas lasts. Blood throbs under the skin of this woman's face, the wind shakes a branch, a group of men prepare to leave. In a novel or a film, the young man will stop smiling, but he will start to smile again when we turn to this page or that moment. Art preserves, and it is the only thing in the world that is preserved. It preserves and is preserved in itself (*quid juris?*), although actually it lasts no longer than its support and materials—stone, canvas, chemical color, and so on (*quid facti?*). The young girl maintains the pose that she has had for five thousand years, a gesture that no longer depends on whoever made it. The air still has the turbulence, the gust of wind, and the light that it had that day last year, and it no longer depends on whoever was breathing it that morning. If art preserves it does not do so like industry, by adding a substance to make the thing last. The thing became independent of its "model" from the start, but it is also independent of other possible personae who are themselves artists-things, personae of painting breathing this air of painting. And it is no less independent of the viewer or hearer, who only experience it after, if they have the strength for it. What about the creator? It is independent of the creator through the self-positing of the created, which is preserved in itself. What is preserved—the thing or the work of art—is *a bloc of sensations, that is to say, a compound of percepts and affects.*

Percepts are no longer perceptions; they are independent of a state of those who experience them. Affects are no longer feelings or affections; they go beyond the strength of those who undergo them. Sensations, percepts, and affects are *beings* whose validity lies in themselves and exceeds any lived. They could be said to exist in the absence of man because man, as he is caught in stone, on the canvas, or by words, is himself a compound of percepts and affects. The work of art is a being of sensation and nothing else: it exists in itself.

Harmonies are affects. Consonance and dissonance, harmonies of tone or color, are affects of music or painting. Rameau emphasized the identity of harmony and affect. The artist creates blocs of percepts and affects, but the only law of creation is that the compound must stand up on its own. The artist's greatest difficulty is to make it *stand up on its own.* Sometimes this requires what is, from the viewpoint of an implicit model, from the viewpoint of lived perceptions and affections, great geometrical improbability, physical imperfection, and organic abnormality. But these sublime errors accede to the necessity of art if they are internal means of standing up (or sitting or lying). There is a pictorial possibility that has nothing to do with physical possibility and that endows the most acrobatic postures with the sense of balance. On the other hand, many works that claim to be art do not stand up for an instant. Standing up alone does not mean having a top and a bottom or being upright (for even houses are drunk and askew); it is only the act by which the compound of created sensations is preserved in itself—a monument, but one that may be contained in a few marks or a few lines, like a poem by Emily Dickinson. Of the sketch of an old, worn-out ass, "How marvellous! It's done with two strokes, but set on immutable bases," where the sensation bears witness all the more to years of "persistent, tenacious, disdainful work."[1] In music, the minor mode is a test that is especially essential since it sets the musician the challenge of wresting it from its ephemeral combinations in order to make it solid and durable, self-preserving, even in acrobatic positions. The sound must be held no less in its extinction than in its production and development. Through his admiration of Pissaro and Monet, what Cézanne had against the Impressionists was that the optical mixture of colors was not enough to create a compound sufficiently "solid and lasting like the art of the museums," like "the perpetuity of blood" in Rubens.[2] This is a way of speaking, because Cézanne does not add something that would preserve Impressionism; he seeks instead a different solidity, other bases and other blocs.

The question of whether drugs help the artist to create these beings of sensation, whether they are part of art's internal means that really lead us to the "doors of perception" and reveal to us percepts and affects, is given a general answer inasmuch as drug-induced compounds are usually extraordinarily flaky, unable to preserve themselves, and break up as soon as they are made or looked at. We may also admire children's drawings, or rather be moved by them, but they rarely stand up and only resemble Klee or Miró if we do not look at them for long. The paintings of the mad, on the contrary, often hold up, but on condition of being crammed full, with no empty space remaining. However, blocs need pockets of air and emptiness, because even the void is sensation. All sensation is composed with the void in composing itself with itself, and everything holds together on earth and in the air, and preserves the void, is preserved in the void by preserving itself. A canvas may be completely full to the point that even the air no longer gets through, but it is only a work of art if, as the Chinese painter says, it nonetheless saves enough empty space for horses to prance in (even if this is only through the variety of planes).[3]

We paint, sculpt, compose, and write with sensations. We paint, sculpt, compose, and write sensations. As percepts, sensations are not perceptions referring to an object (reference): if they resemble something it is with a resemblance produced with their own methods; and the smile on the canvas is made solely with colors, lines, shadow, and light. If resemblance haunts the work of art, it is because sensation refers only to its material: it is the percept or affect of the material itself, the smile of oil, the gesture of fired clay, the thrust of metal, the crouch of Romanesque stone, and the ascent of Gothic stone. The material is so varied in each case (canvas support, paintbrush or equivalent agent, color in the tube) that it is difficult to say where in fact the material ends and sensation begins; preparation of the canvas, the track of the brush's hair, and many other things besides are obviously part of the sensation. How could the sensation be preserved without a material capable of lasting? And however short the time it lasts, this time is considered as a duration. We will see how the plane of the material ascends irresistibly and invades the plane of composition of the sensations themselves to the point of being part of them or indiscernible from them. It is in this sense that the painter is said to be a painter and nothing but a painter, "with color seized as if just pressed out of the tube, with the imprint of each hair of his brush," with this blue that is not a water blue "but a liquid paint blue." And yet, in principle at least, sensation is not the same thing as the material. What is preserved by right is not the material, which constitutes only the de facto condition, but, insofar as this condition is satisfied (that is, that canvas, color, or stone does not crumble into dust), it is the percept or affect that is preserved in itself. Even if the material lasts for only a few seconds it will give sensation the power to exist and be preserved in itself *in the eternity that coexists with this short duration.* So long as the material lasts, the sensation enjoys an eternity in those very moments. Sensation is not realized in the material without the material passing completely into the sensation, into the percept or affect. All the material becomes expressive. It is the affect that is metallic, crystalline, stony, and so on; and the sensation is not colored but, as Cézanne said, coloring. That is why those who are nothing but painters are also more than painters, because they "bring before us, in front of the fixed canvas," not the resemblance but the pure sensation "of a tortured flower, of a landscape slashed, pressed, and plowed," giving back "the water of the painting to nature."[4] One material is exchanged for another, like the violin for the piano, one kind of brush for another, oil for pastel, only inasmuch as the compound of sensations requires it. And, however strong an artist's interest in science, a compound of sensations will never be mistaken for the "mixtures" of material that science determines in states of affairs, as is clearly shown by the "optical mixture" of the impressionists.

By means of the material, the aim of art is to wrest the percept from perceptions of objects and the states of a perceiving subject, to wrest the affect from affections as the transition from one state to another: to extract a bloc of sensations, a pure being of sensations. A method is needed, and this varies with every artist and forms part of the work: we need only compare Proust and Pessoa, who invent different procedures in the search for the sensation as being.[5] In this respect the writer's position is no different from that of the painter, musician, or architect. The writer's specific materials are words and syntax, the created syntax that ascends irresistibly into his work and passes into sensation. Memory, which summons forth only old perceptions, is obviously not enough to get away from lived perceptions; neither is an involuntary memory that adds reminiscence as the present's preserving factor. Memory plays a small part in art (even and especially in Proust). It is true that every work of art is a *monument,* but here the monument is not something commemorating a past, it is a bloc of present sensations that owe their preservation only to themselves and that provide the event with the compound that celebrates it. The monument's action is not memory but fabulation. We write not with childhood memories but through blocs of childhood that are the becoming-child of the present. Music is full of them. It is not memory that is needed but a complex material that is found not in memory but in words and sounds: "Memory, I hate you." We attain to the percept and the affect only as to autonomous and sufficient beings that no longer owe anything to those who experience or have experienced them: Combray like it never was, is, or will be lived; Combray as cathedral or monument.

If methods are very different, not only in the different arts but in different artists, we can nevertheless characterize some great monumental types, or "varieties," of compounds of sensations: *the vibration,* which characterizes the simple sensation (but it is already durable or compound, because it rises and falls, implies a constitutive difference of level, follows an invisible thread that is more nervous than cerebral); *the embrace or the clinch* (when two sensations resonate in each other by embracing each other so tightly in a clinch of what are no more than "energies"); *withdrawal, division, distension* (when, on the contrary, two sensations draw apart, release themselves, but so as now to be brought together by the light, the air, or the void that sinks between them or into them, like a wedge that is at once so dense and so light that it extends in every direction as the distance grows, and forms a bloc that no longer needs a support). Vibrating sensation—coupling sensation—opening or splitting, hollowing out sensation. These types are displayed almost in their pure state in sculpture, with its sensations of stone, marble, or metal, which vibrate according to the order of strong and weak beats, projections and hollows, its powerful clinches that intertwine them, its development of large spaces between groups or within a single group where we no longer know whether it is the light or the air that sculpts or is sculpted.

The novel has often risen to the percept—not perception of the moor in Hardy but the moor as percept; oceanic percepts in Melville;

urban percepts, or those of the mirror, in Virginia Woolf. The landscape *sees*. Generally speaking, what great writer has not been able
to create these beings of sensation, which preserve in themselves the
hour of a day, a moment's degree of warmth (Faulkner's hills, Tol-
stoy's or Chekhov's steppes)? The percept is the landscape before
man, in the absence of man. But why do we say this, since in all these
cases the landscape is not independent of the supposed perceptions of
the characters and, through them, of the author's perceptions and
memories? How could the town exist without or before man, or the
mirror without the old woman it reflects, even if she does not look at
herself in it? This is Cézanne's enigma, which has often been com-
mented upon: "Man absent from but entirely within the landscape."
Characters can only exist, and the author can only create them,
because they do not perceive but have passed into the landscape and
are themselves part of the compound of sensations. Ahab really does
have perceptions of the sea, but only because he has entered into a
relationship with Moby Dick that makes him a becoming-whale and
forms a compound of sensations that no longer needs anyone: ocean.
It is Mrs. Dalloway who perceives the town—but because she has
passed into the town like "a knife through everything" and becomes
imperceptible herself. *Affects are precisely these nonhuman becomings
of man,* just as percepts—including the town—are *nonhuman landscapes of nature*. Not a "minute of the world passes," says Cézanne,
that we will preserve if we do not "become that minute."[6] We
are not in the world, we become with the world; we become by
contemplating it. Everything is vision, becoming. We become universes. Becoming animal, plant, molecular, becoming zero. Kleist is
no doubt the author who most wrote with affects, using them like
stones or weapons, seizing them in becomings of sudden petrification
or infinite acceleration, in the becoming-bitch of Penthesilea and
her hallucinated percepts. This is true of all the arts: what strange
becomings unleash music across its "melodic landscapes" and its
"rhythmic characters," as Messiaen says, by combining the molecular

and the cosmic, stars, atoms, and birds in the same being of sensation? What terror haunts Van Gogh's head, caught in a becoming-sunflower? In each case style is needed—the writer's syntax, the musician's modes and rhythms, the painter's lines and colors—to raise lived perceptions to the percept and lived affections to the affect.

We dwell on the art of the novel because it is the source of a misunderstanding: many people think that novels can be created with our perceptions and affections, our memories and archives, our travels and fantasies, our children and parents, with the interesting characters we have met and, above all, the interesting character who is inevitably oneself (who isn't interesting?), and finally with our opinions holding it all together. If need be, we can invoke great authors who have done nothing but recount their lives—Thomas Wolfe or Henry Miller. Generally we get composite works in which we move about a great deal but in search of a father who is found only in ourself: the journalist's novel. We are not spared the least detail, in the absence of any really artistic work. The cruelty we may have seen and the despair we have experienced do not need to be transformed a great deal in order to produce yet again the opinion that generally emerges about the difficulties of communication. Rossellini saw this as a reason for giving up art: art was allowing itself to be invaded too much by infantilism and cruelty, both cruel and doleful, whining and satisfied at the same time, so that it was better to abandon it.[7] More interestingly, Rosselini saw the same thing taking place in painting. But it is literature primarily that has constantly maintained an equivocal relationship with the lived. We may well have great powers of observation and much imagination, but is it possible to write with perceptions, affections, and opinions? Even in the least autobiographical novels we see the confrontation and intersection of the opinions of a multitude of characters, all in accordance with the perceptions and affections of each character with his social situation and individual adventures, and all of it swept up in the vast current of the author's opinion, which, however, divides itself so as to rebound on the characters, or which hides itself so that readers can form their own: this is indeed how Bakhtin's great theory of the novel begins (happily it does not end there; it is precisely the "parodic" basis of the novel).

Creative fabulation has nothing to do with a memory, however exaggerated, or with a fantasy. In fact, the artist, including the novelist, goes beyond the perceptual states and affective transitions of the lived. The artist is a seer, a becomer. How would he recount what happened to him, or what he imagines, since he is a shadow? He has seen something in life that is too great, too unbearable also, and the mutual embrace of life with what threatens it, so that the corner of nature or districts of the town that he sees, along with their characters, accede to a vision that, through them, composes the percepts of that life, of that moment, shattering lived perceptions into a sort of cubism, a sort of simultaneism, of harsh or crepuscular light, of purple or blue, which have no other object or subject than themselves. "What we call styles," said Giacometti, "are those visions fixed in time and space." It is always a question of freeing life wherever it is imprisoned, or of tempting it into an uncertain combat. The death of the porcupine in Lawrence and the death of the mole in Kafka are almost unbearable acts of the novelist. Sometimes it is necessary to lie down on the earth, like the painter does also, in order to get to the "motif," that is to say, the percept. Percepts can be telescopic or microscopic, giving characters and landscapes giant dimensions as if they were swollen by a life that no lived perception can attain. Balzac's greatness. It is of little importance whether these characters *are* mediocre: they *become* giants, like Bouvard and Pecuchet, Bloom and Molly, Mercier and Camier, without ceasing to be what they are. It is by dint of mediocrity, even of stupidity or infamy, that they are able to become not simple (they are never simple) but gigantic. Even dwarves and cripples will do: all fabulation is the fabrication of giants.[8] Whether mediocre or grandiose, they are too alive to be livable or lived. Thomas Wolfe extracts a giant from his father, and Henry Miller extracts a dark planet from the city. Wolfe may describe the people of old Catawba through their stupid opinions and their mania for discussion, but what he does is set up the secret monument of their solitude, their desert, their eternal earth, and their forgotten, unnoticed lives. Faulkner may also cry out: oh, men of Yoknapatawpha. It is said that the monumental novelist is himself "inspired" by the lived, and this is true: M. de Charlus closely resembles Montesquiou, but between Montesquiou and M. de Charlus there is ultimately roughly the same relationship as between the barking animal-dog and the celestial constellation-Dog.

How can a moment of the world be rendered durable or made to exist by itself? Virginia Woolf provides an answer that is as valid for painting and music as it is for writing: "Saturate every atom," "eliminate all waste, deadness, superfluity," everything that adheres to our current and lived perceptions, everything that nourishes the mediocre novelist; and keep only the saturation that gives us the percept. "It must include nonsense, fact, sordidity: *but made transparent*"; "I want to put practically everything in; yet to saturate."[9] Through having reached the percept as "the sacred source," through having seen Life in the living or the Living in the lived, the novelist or painter returns breathless and with bloodshot eyes. They are athletes—not athletes who train their bodies and cultivate the lived, no matter how many writers have succumbed to the idea of sport as a way of heightening art and life, but bizarre athletes of the "fasting-artist" type, or the "great Swimmer" who does not know how to swim. It is not an organic or muscular athleticism but its inorganic double, "an affective Athleticism," an athleticism of becoming that reveals only forces that are not its own—"plastic specter."[10] In this respect artists are like philosophers. What little health they possess is often too fragile, not because of their illnesses or neuroses but because they have seen something in life that is too much for anyone, too much for themselves, and that has put on them the quiet mark of death. But this something is also the source or breath that supports them through the illnesses of the lived (what Nietzsche called health). "Perhaps one day we will know that there wasn't any art but only medicine."[11]

The affect goes beyond affections no less than the percept goes beyond perceptions. The affect is not the passage from one lived state to another but man's nonhuman becoming. Ahab does not imitate Moby Dick, and Penthesilea does not "act" the bitch: becoming is neither an imitation nor an experienced sympathy, nor even an imaginary identification. It is not resemblance, although there is resemblance. But it is only a produced resemblance. Rather, becoming is an extreme contiguity within a coupling of two sensations without resemblance or, on the contrary, in the distance of a light that captures both of them in a single reflection. André Dhotel knew how to place his characters in strange plant-becomings, becoming tree or aster: this is not the transformation of one into the other, he says, but something passing from one to the other.[12] This something can be specified only as sensation. It is a zone of indetermination, of indiscernibility, as if things, beasts, and persons (Ahab and Moby Dick, Penthesilea and the bitch) endlessly reach that point that immediately precedes their natural differentiation. This is what is called an *affect*. In *Pierre; or, The Ambiguities,* Pierre reaches the zone in which he can no longer distinguish himself from his half-sister, Isabelle, and he becomes woman. Life alone creates such zones where living beings whirl around, and only art can reach and penetrate them in its enterprise of co-creation. This is because from the moment that the material passes into sensation, as in a Rodin sculpture, art itself lives on these zones of indetermination. They are blocs. Painting needs more than the skill of the draftsman who notes resemblances between human and animal forms and gets us to witness their transformation: on the contrary, it needs the power of a ground that can dissolve forms and impose the existence of a zone in which we no longer know which is animal and which human, because something like the triumph or monument of their nondistinction rises up—as in Goya or even Daumier or Redon. The artist must create the syntacti-

cal or plastic methods and materials necessary for such a great undertaking, which re-creates everywhere the primitive swamps of life
(Goya's use of etching and aquatint). The affect certainly does not
undertake a return to origins, as if beneath civilization we would
rediscover, in terms of resemblance, the persistence of a bestial or
primitive humanity. It is within our civilization's temperate surroundings that equatorial or glacial zones, which avoid the differentiation of genus, sex, orders, and kingdoms, currently function and
prosper. It is a question only of ourselves, here and now; but what is
animal, vegetable, mineral, or human in us is now indistinct—even
though we ourselves will especially acquire distinction. The maximum determination comes from this bloc of neighborhood like a
flash.

It is precisely because opinions are functions of lived experience
that they claim to have a certain knowledge of affections. Opinions
prevail on human passions and their eternity. But, as Bergson ob-
served, one has the impression that opinion misjudges affective states
and groups them together or separates them wrongly.[13] It is not even
enough to do what psychoanalysis does and give forbidden objects to
itemized affections or substitute simple ambivalences for zones of
indetermination. A great novelist is above all an artist who invents
unknown or unrecognized affects and brings them to light as the
becoming of his characters: the crepuscular states of knights in the
novels of Chrétien de Troyes (in relation to a possible concept of
chivalry), the states of almost catatonic "rest" that merge with duty
according to Mme de Lafayette (in relation to a concept of quietism),
on up to Beckett's state, as affects that are all the more imposing as
they are poor in affections. When Zola suggests to his readers, "take
note; my characters do not suffer from remorse," we should see not
the expression of a physiologist's thesis but the ascription of new
affects that arise with the creation of characters in naturalism: the
Mediocre, the Pervert, the Beast (and what Zola calls instinct is

inseparable from a becoming-animal). When Emily Brontë traces the bond between Heathcliff and Catherine, she invents a violent affect, like a kinship between two wolves, which above all should not be mistaken for love. When Proust seems to be describing jealousy in such minute detail, he is inventing an affect, because he constantly reverses the order in affections presupposed by opinion, according to which jealousy would be an unhappy consequence of love: for him, on the contrary, jealousy is finality, destination; and if we must love, it is so that we can be jealous, jealousy being the meaning of signs—affect as semiology. When Claude Simon describes the incredible passive love of the earth-woman, he sculpts an affect of clay. He may say, "this is my mother," and we believe him since he says it, but it is a mother who has passed into sensation and to whom he erects a monument so original that she no longer has an ascribable relationship with her real son but, more distantly, with another created character, Faulkner's Eula. It is in this way that, from one writer to another, great creative affects can link up or diverge, within compounds of sensations that transform themselves, vibrate, couple, or split apart: it is these beings of sensation that account for the artist's relationship with a public, for the relation between different works by the same artist, or even for a possible affinity between artists.[14] The artist is always adding new varieties to the world. Beings of sensation are varieties, just as the concept's beings are variations, and the function's beings are variables.

It should be said of all art that, in relation to the percepts or visions they give us, artists are presenters of affects, the inventors and creators of affects. They not only create them in their work, they give them to us and make us become with them, they draw us into the compound. Van Gogh's sunflowers are becomings, like Dürer's thistles or Bonnard's mimosas. Redon entitled a lithograph "There was perhaps a first vision attempted in the flower." The flower sees—pure and simple terror: "And do you see that sunflower looking in through the bedroom window? It stares into my room all day."[15] A floral history of painting is like the endlessly and continuously resumed creation of the percepts and affects of flowers. Whether through words, colors, sounds, or stone, art is the language of sensations. Art does not have opinions. Art undoes the triple organization of perceptions, affections, and opinions in order to substitute a monument composed of percepts, affects, and blocs of sensations that take the place of language. The writer uses words, but by creating a syntax that makes them pass into sensation that makes the standard language stammer, tremble, cry, or even sing: this is the style, the "tone," the language of sensations, or the foreign language within language that summons forth a people to come, "Oh, people of old Catawba," "Oh, people of Yoknapatawpha." The writer twists language, makes it vibrate, seizes hold of it, and rends it in order to wrest the percept from perceptions, the affect from affections, the sensation from opinion—in view, one hopes, of that still-missing people. "I repeat—my memory is not loving but inimical, and it labors not to reproduce but to distance the past. What was it my family wished to say? I do not know. It was tongue-tied from birth—but it had, nevertheless, something that it might have said. Over my head and over the head of many of my contemporaries there hangs the congenital tongue-tie. We were not taught to speak but to babble—and only by listening to the swelling noise of the age and bleached by the foam on the crest of its wave did we acquire a language."[16] This is, precisely, the task of all art and, from colors and sounds, both music and painting similarly extract new harmonies, new plastic or melodic landscapes, and new rhythmic characters that raise them to the height of the earth's song and the cry of humanity: that which constitutes tone, health, becoming, a visual and sonorous bloc. A monument does not commemorate or celebrate something that happened but confides to the ear of the future the persistent sensations that embody the event: the constantly renewed suffering of men and women, their re-created protestations, their constantly resumed struggle. Will this all be in vain because suffering is eternal and revolutions do not survive their victory? But the success of a revolution resides only in itself, precisely in the vibrations, clinches, and openings it gave to men and women at the moment of its making and that composes in itself a monument that is always in the process of becoming, like those tumuli to which each new traveler adds a stone. The victory of a revolution is immanent and consists in the new bonds it installs between people, even if these bonds last no longer than the revolution's fused material and quickly give way to division and betrayal.

Aesthetic figures, and the style that creates them, have nothing to do with rhetoric. They are sensations: percepts and affects, landscapes and faces, visions and becomings. But is not the philosophical concept defined by becoming, and almost in the same terms? Still, aesthetic figures are not the same as conceptual personae. It may be that they pass into one another, in either direction, like Igitur or Zarathustra, but this is insofar as there are sensations of concepts and concepts of sensations. It is not the same becoming. Sensory becoming is the action by which something or someone is ceaselessly becoming-other (while continuing to be what they are), sunflower or Ahab, whereas conceptual becoming is the action by which the common event itself eludes what is. Conceptual becoming is heterogeneity grasped in an absolute form; sensory becoming is otherness caught in a matter of expression. The monument does not actualize the virtual event but incorporates or embodies it: it gives it a body, a life, a universe. This was how Proust defined the art-monument by that life higher than the "lived," by its "qualitative differences," its "universes" that construct their own limits, their distances and proximities, their constellations and the blocs of sensations they put into motion—Rembrandt-universe or Debussy-universe. These universes are neither virtual nor actual; they are possibles, the possible as aesthetic category ("the possible or I shall suffocate"), the existence of the possible, whereas events are the reality of the virtual, forms of a thought-Nature that survey every possible universe. This is not to say that the concept precedes sensation in principle: even a concept of sensation must be created with its own means, and a sensation exists in its possible universe without the concept necessarily existing in its absolute form.

Can sensation be assimilated to an original opinion, to *Urdoxa* as the world's foundation or immutable basis? Phenomenology finds sensation in perceptual and affective "a priori materials" that transcend the perceptions and affections of the lived: Van Gogh's yellow or Cézanne's innate sensations. As we have seen, phenomenology must become the phenomenology of art because the immanence of the lived to a transcendental subject must be expressed in transcendent functions that not only determine experience in general but traverse the lived itself here and now, and are embodied in it by constituting living sensations. The being of sensation, the bloc of percept and affect, will appear as the unity or reversibility of feeling and felt, their intimate intermingling like hands clasped together: it is the *flesh* that, at the same time, is freed from the lived body, the perceived world, and the intentionality of one toward the other that is still too tied to experience; whereas flesh gives us the being of sensation and bears the original opinion distinct from the judgment of experience—flesh of the world and flesh of the body that are exchanged as correlates, ideal coincidence.[17] A curious Fleshism inspires this final avatar of phenomenology and plunges it into the mystery of the incarnation. It is both a pious and a sensual notion, a mixture of sensuality and religion, without which, perhaps, flesh could not stand up by itself (it would slide down the bones, as in Bacon's figures). The question of whether flesh is adequate to art can be put in this way: can it support percept and affect, can it constitute the being of sensation, or must it not itself be supported and pass into other powers of life?

Flesh is not sensation, although it is involved in revealing it. We spoke too quickly when we said that sensation embodies. Sometimes flesh is painted with pink (superimpositions of red and white), and

sometimes with broken tones [*tons rompus**], a juxtaposition of complementaries in unequal proportions. But what constitutes sensation
is the becoming animal or plant, which wells up like a flayed beast or
peeled fruit beneath the bands of pink in the most graceful, delicate
nude, Venus in the mirror; or which suddenly emerges in the fusion,
firing, or casting of broken tones, like the zone of indiscernibility of
beast and man. Perhaps it would be an interference or chaos, were
there not a second element to make the flesh hold fast. Flesh is only
the thermometer of a becoming. The flesh is too tender. The second
element is not so much bone or skeletal structure as house or framework. The body blossoms in the house (or an equivalent, a spring, a
grove). Now, what defines the house are "sections," that is to say,
the pieces of differently oriented planes that provide flesh with its
framework—foreground and background, horizontal and vertical sections, left and right, straight and oblique, rectilinear or curved.[18]
These sections are walls but also floors, doors, windows, French
windows, and mirrors, which give sensation the power to stand on
its own within autonomous *frames*. They are the sides of the bloc of
sensation. There are certainly two signs of the genius of great painters, as well as of their humility: the respect, almost dread, with which
they approach and enter into color; and the care with which they join
together the sections or planes on which the type of depth depends.
Without this respect and care painting is nothing, lacking work and
thought. The difficult part is not to join hands but to join planes—to
produce bulging with joined planes or, on the contrary, to break
them open or cut them off. The two problems, the architecture of
planes and the regime of color, are often mixed up. As for the joining
of horizontal and vertical planes in Cézanne, "Planes in color, planes!

*There does not seem to be a standard equivalent technical term in English for the
French *tons rompus,* which means colors or tones made up of several different colors or
tones. Van Gogh's letters, which are a principal reference point for this notion, speak
of colors that are "broken" with other colors; following this we have translated the
term as "broken tones."

The colored place where the heart of the planes is fused." No two great painters, or even oeuvres, work in the same way. However there are tendencies in a painter: in Giacometti, for example, the receding horizontal planes differ from right to left and seem to come together on the thing (the flesh of the small apple), but like a pincer that would pull it backward and make it disappear if a vertical plane, of which we see only the thread without thickness, did not fix it, checking it at the last moment, giving it a durable existence, in the form of a long pin passing through it and rendering it spindly in turn. The house takes part in an entire becoming. It is life, the "nonorganic life of things." In every way possible, the house-sensation is defined by the joining of planes in accordance with a thousand orientations. The house itself (or its equivalent) is the finite junction of colored planes.

The third element is the universe, the cosmos. Not only does the open house communicate with the landscape, through a window or a mirror, but the most shut-up house opens onto a universe. Monet's house finds itself endlessly caught up by the plant forces of an unrestrained garden, a cosmos of roses. A universe-cosmos is not flesh. Neither is it sections, joined up parts of planes, or differently oriented planes, although it may be constituted by the connection of every plane to infinity. But ultimately the universe appears as the area of plain, uniform color [*l'aplat**], the single great plane, the colored void, the monochrome infinite. The French window, as in Matisse, now opens only onto an area of plain, uniform black. The flesh, or rather the figure, is no longer the inhabitant of the place, of the house, but of the universe that supports the house (becoming). *It is like a passage from the finite to the infinite,* but also from territory to

---

*As with *tons rompus*, the term with which it is contrasted here, there does not seem to be a standard English equivalent for the French *aplat*. The noun has connotations of flatness, following the verb *aplatir* (to flatten or smooth out), but in painting it signifies areas of plain, uniform color. In the absence of a single English word we have decided to use the entire phrase "area of plain, uniform color."

deterritorialization. It is indeed the moment of the infinite: infinitely varied infinites. In Van Gogh, Gauguin, or, today, Bacon, we see the immediate tension between flesh and the area of plain, uniform color surging forth, between the flows of broken tones [*tons rompus*] and the infinite band of a pure, homogeneous, vivid, and saturated color ("instead of painting the ordinary wall of the mean room, I paint infinity, a plain background of the richest, intensest blue").[19] It is true that the monochrome area of plain color is something other than a background. And when painting wants to start again at zero, by constructing the percept as a minimum before the void, or by bringing it closer to the maximum of the concept, it works with monochrome freed from any house or flesh. Blue in particular takes on the infinite and turns the percept into a "cosmic sensibility" or into that which is most conceptual or "propositional" in nature—color in the absence of man, man who has passed into color. But if the blue (or black or white) is exactly the same within a picture, or from one picture to another, then it is the painter who becomes blue—"Yves the monochrome"—in accordance with a pure affect that topples the universe into the void and leaves the painter above all with nothing to do.[20]

The colored or, rather, coloring void, is already force. Most of the great monochromes of modern painting no longer need to resort to little mural bouquets but present subtle imperceptible variations (which are constitutive of a percept nevertheless), either because they are cut off or edged on one side by a band, ribbon, or section of a different color or tone that, through proximity or distance, changes the intensity of the area of plain, uniform color or because they present almost virtual linear or circular figures, in matching tones, or because they are holed or slit: these are problems of junction, once again, but considerably expanded. In short, the area of plain, uniform color vibrates, clenches or cracks open because it is the bearer of glimpsed forces. And this, first of all, is what makes painting abstract: summoning forces, populating the area of plain, uniform color with the forces it bears, making the invisible forces visible in themselves, drawing up figures with a geometrical appearance but that are no more than forces—the forces of gravity, heaviness, rotation, the vortex, explosion, expansion, germination, and time (as music may be said to make the sonorous force of time audible, in Messiaen for example, or literature, with Proust, to make the illegible force of time legible and conceivable). Is this not the definition of the percept itself—to make perceptible the imperceptible forces that populate the world, affect us, and make us become? Mondrian achieves this by simple differences between the sides of a square, Kandinsky by linear "tensions," and Kupka by planes curved around the point. From the depths of time there comes to us what Worringer called the abstract and infinite northern line, the line of the universe that forms ribbons, strips, wheels, and turbines, an entire "vitalized geometry," *rising to the intuition of mechanical forces,* constituting a powerful nonorganic life.[21] Painting's eternal object is this: to paint forces, like Tintoretto.

Perhaps also we rediscover the house and the body?—because the infinite area of plain, uniform color is often that onto which the window or door opens; or it is the wall of the house itself, or the floor. Van Gogh and Gauguin sprinkle the area of plain, uniform color with little bunches of flowers so as to turn it into wallpaper on which the face stands out in broken tones. In fact, the house does not shelter us from cosmic forces; at most it filters and selects them. Sometimes it turns them into benevolent forces: Archimedes' force, the force of the water's pressure on a graceful body floating in the bath of the house, has never been made visible in painting in the way that Bonnard succeeded in doing in *Le Nu au bain.* But equally, the most baleful forces can come in through the half-open or closed door: cosmic forces themselves are what produce zones of indiscernibility in the broken tones of a face, slapping, scratching, and melting it in every way, and these zones of indiscernibility reveal the forces lurking in the area of plain, uniform color (Bacon). The clinch of forces as percepts and becomings as affects are completely complementary.

According to Worringer, the abstract line of force is rich in animal motifs. Animal, plant, and molecular becomings correspond to cosmic or cosmogenetic forces: to the point that the body disappears into the plain color or becomes part of the wall or, conversely, the plain color buckles and whirls around in the body's zone of indiscernibility. In short, the being of sensation is not the flesh but the compound of nonhuman forces of the cosmos, of man's nonhuman becomings, and of the ambiguous house that exchanges and adjusts them, makes them whirl around like winds. Flesh is only the developer which disappears in what it develops: the compound of sensation. Like all painting, abstract painting is sensation, nothing but sensation. In Mondrian the room accedes to the being of sensation by dividing the infinite empty plane by colored sections that, in turn, give it an infinite openness.[22] In Kandinsky, houses are sources of abstraction that consist less in geometrical figures than in dynamic trajectories and errant lines, "paths that go for a walk" in the surroundings. In Kupka it is first of all on the body that the painter cuts out colored ribbons or sections that will give, in the void, the curved planes that populate it by becoming cosmogenetic sensations. Is sensation spiritual, or already a living concept—the room, house, universe? Abstract art, and then conceptual art, directly pose the question that haunts all painting—that of its relation to the concept and the function.

Perhaps art begins with the animal, at least with the animal that carves out a territory and constructs a house (both are correlative, or even one and the same, in what is called a habitat). The territory-house system transforms a number of organic functions—sexuality, procreation, aggression, feeding. But this transformation does not explain the appearance of the territory and the house; rather it is the other way around: the territory implies the emergence of pure sensory qualities, of sensibilia that cease to be merely functional and become expressive features, making possible a transformation of functions.[23] No doubt this expressiveness is already diffused in life, and the simple field of lilies might be said to celebrate the glory of the skies. But with the territory and the house it becomes constructive and erects ritual monuments of an animal mass that celebrates qualities before extracting new causalities and finalities from them. This emergence of pure sensory qualities is already art, not only in the treatment of external materials but in the body's postures and colors, in the songs and cries that mark out the territory. It is an outpouring of features, colors, and sounds that are inseparable insofar as they become expressive (philosophical concept of territory). Every morning the *Sceno-poetes dentirostris,* a bird of the Australian rain forests, cuts leaves, makes them fall to the ground, and turns them over so that the paler, internal side contrasts with the earth. In this way it constructs a stage for itself like a ready-made; and directly above, on a creeper or a branch, while fluffing out the feathers beneath its beak to reveal their yellow roots, it sings a complex song made up from its own notes and, at intervals, those of other birds that it imitates: it is a complete artist.[24] This is not synesthesia in the flesh but blocs of sensations in the territory—colors, postures, and sounds that sketch out a total work of art. These sonorous blocs are refrains; but there are also refrains of posture and color, and postures and colors are always being introduced into refrains: bowing low, straightening up, dancing in a circle and lines of colors. The whole of the refrain is the being of sensation. Monuments are refrains. In this respect art is continually haunted by the animal. Kafka's art is the most profound meditation on the territory and the house, the burrow, portrait-postures (the inhabitant's lowered head with chin sunk into their chest or, on the contrary, "Shamefaced Lacky" whose angular head goes right through the ceiling); sounds-music (dogs who are musicians in their very postures; Josephine, the singing mouse, of whom it will never be known whether she sings; Gregor whose squeaking combines with his sister's violin in a complex bedroom-house-territory relationship). All that is needed to produce art is here: a house, some postures, colors, and songs—on condition that it all opens onto and launches itself on a mad vector as on a witch's broom, a line of the universe or of deterritorialization—*Perspective on a Room with Occupants* (Klee).

Every territory, every habitat, joins up not only its spatiotemporal but its qualitative planes or sections: a posture and a song for example, a song and a color, percepts and affects. And every territory encompasses or cuts across the territories of other species, or intercepts the trajectories of animals without territories, forming interspecies junction points. It is in this sense that, to start with, Uexkühl develops a melodic, polyphonic, and contrapuntal conception of Nature. Not only does birdsong have its own relationships of counterpoint but it can find these relationships in the song of other species, and it may even imitate these other songs as if it were a question of occupying a maximum of frequencies. The spider's web contains "a very subtle portrait of the fly," which serves as its counterpoint. On the death of the mollusk, the shell that serves as its house becomes the counterpoint of the hermit crab that turns it into its own habitat, thanks to its tail, which is not for swimming but is prehensile, enabling it to capture the empty shell. The tick is organically constructed in such a way that it finds its counterpoint in any mammal whatever that passes below its branch, as oak leaves arranged in the form of tiles find their counterpoint in the raindrops that stream over them. This is not a teleological conception but a melodic one in which we no longer know what is art and what nature ("natural technique"). There is counterpoint whenever a melody arises as a "motif" within another melody, as in the marriage of bumblebee and snapdragon. These relationships of counterpoint join planes together, form compounds of sensations and blocs, and determine becomings. But it is not just these determinate *melodic compounds,* however generalized, that constitute nature; another aspect, an infinite *symphonic plane of composition,* is also required: from House to universe. From endosensation to exosensation. This is because the territory does not merely isolate and join but opens onto cosmic forces that arise from within or come from outside, and renders their effect on the inhabitant perceptible. The oak's plane of composition is what supports or includes the force of the acorn's development and the force of formation of raindrops, and the tick's plane of composition is what supports the force of light, which can attract the insect to the end of a branch to a sufficient height, and the force of weight with which it lets itself fall onto the passing mammal—and between them nothing, an alarming void that can last for years if no mammals pass by.[25] Sometimes forces blend into one another in subtle transitions, decompose hardly glimpsed; and sometimes they alternate or conflict with one another. Sometimes they allow themselves to be selected by the territory, and the most benevolent ones are those that enter the house. Sometimes they send out a mysterious call that draws the inhabitant from the territory and launches it on an irresistible voyage, like chaffinches that suddenly assemble in their millions or crayfish that set off in step on an immense pilgrimage to the bottom of the water. Sometimes they swoop down on the territory, turn it upside down, wickedly, restoring the chaos from which, with difficulty, the territory came. But if nature is like art, this is always because it combines these two living elements in every way: House and Universe, *Heimlich* and *Unheimlich*, territory and deterritorialization, finite melodic compounds and the great infinite plane of composition, the small and large refrain.

Art begins not with flesh but with the house. That is why architecture is the first of the arts. When Dubuffet tries to identify a certain condition of *art brut*, he turns first of all to the house, and all his work stands between architecture, sculpture, and painting. And, not going beyond form, the most scientific architecture endlessly produces and joins up planes and sections. That is why it can be defined by the "frame," by an interlocking of differently oriented frames, which will be imposed on the other arts, from painting to the cinema. The prehistory of the picture has been presented as passing through the fresco within the frame of the wall, stained glass within the frame of the window, and mosaic within the frame of the floor: "The frame is the umbilicus that attaches the picture to the monument of which it is the reduction," like the gothic frame, with small columns, diagonal ribs, and openwork spire.[26] By making architecture the first art of the frame, Bernard Cache is able to list a certain number of enframing forms that do not determine in advance any concrete content or function of the edifice: the wall that cuts off, the window that captures or selects (in direct contact with the territory), the ground-floor that wards off or rarefies ("rarefying the earth's relief so as to give a free path to human trajectories"), the roof that envelops the place's singularity ("the sloping roof puts the edifice on a hill"). Interlocking these frames or joining up all these planes—wall section, window section, floor section, slope section—is a composite system rich in points and counterpoints. The frames and their joins hold the compounds of sensations, hold up figures, and intermingle with their upholding, with their own appearance. These are the faces of a dice of sensation. Frames or sections are not coordinates; they belong to compounds of sensations whose faces, whose interfaces, they constitute. But however extendable this system may be, it still needs a vast plane of composition that carries out a kind of *deframing* following lines of flight that pass through the territory only in order to open it onto the universe, that go from house-territory to town-cosmos, and that now dissolve the identity of the place through variation of the earth, a town having not so much a place as vectors folding the abstract line of relief. On this plane of composition, as on "an abstract vectorial space," geometrical figures are laid out—cone, prism, dihedron, simple plane—which are no more than cosmic forces capable of merging, being transformed, confronting each other, and alternating; world before man yet produced by man.[27] The planes must now be taken apart in order to relate them to their intervals rather than to one another and in order to create new affects.[28] We have seen that painting pursued the same movement. The frame or the picture's edge is, in the first place, the external envelope of a series of frames or sections that join up by carrying out counterpoints of lines and colors, by determining compounds of sensations. But the picture is also traversed by a deframing power that opens it onto a plane of composition or an infinite field of forces. These processes may be very diverse, even at the level of the external frame: irregular forms, sides that do not meet, Seurat's painted or stippled frames, and Mondrian's squares standing on a corner, all of which give the picture the power to leave the canvas. The painter's action never stays within the frame; it leaves the frame and does not begin with it.

Literature, and especially the novel, seems to be in the same situation. What matters is not, as in bad novels, the opinions held by characters in accordance with their social type and characteristics but rather the relations of counterpoint into which they enter and the compounds of sensations that these characters either themselves experience or make felt in their becomings and their visions. Counterpoint serves not to report real or fictional conversations but to bring out the madness of all conversation and of all dialogue, even interior dialogue. Everything that novelists must extract from the perceptions, affections, and opinions of their psychosocial "models" passes entirely into the percepts and affects to which the character must be raised without holding on to any other life. And this entails a vast plane of composition that is not abstractly preconceived but constructed as the work progresses, opening, mixing, dismantling, and reassembling increasingly unlimited compounds in accordance with the penetration of cosmic forces. Bakhtin's theory of the novel goes in this direction by showing, from Rabelais to Dostoyevsky, the coexistence of contrapuntal, polyphonic, and plurivocal compounds with an architectonic or symphonic plane of composition.[29] A novelist like Dos Passos achieves an extraordinary art of counterpoint in the compounds he forms with characters, current events, biographies, and camera eyes, at the same time as a plane of composition is expanded to infinity so as to sweep everything up into Life, into Death, the town cosmos. If we return to Proust, it is because he more than anyone else made the two elements, although present in each other, almost follow one another; the plane of composition, for life and for death, emerges gradually from compounds of sensation that he draws up in the course of lost time, until appearing in itself with time regained, the force, or rather the forces, of pure time that have now become perceptible. Everything begins with Houses, each of which must join up its sections and hold up compounds—Combray, the Guermantes' house, the Verdurins' salon—and the houses are themselves joined together according to interfaces, but a planetary Cosmos is already there, visible through the telescope, which ruins or transforms them and absorbs them into an infinity of the patch of uniform color. Everything begins with refrains, each of which, like the little phrase of Vinteuil's sonata, is composed not only in itself but with other, variable sensations, like that of an unknown passer-by, like Odette's face, like the leaves of the Bois de Boulogne—and everything comes to an end at infinity in the great Refrain, the phrase of the septet in perpetual metamorphosis, the song of the universe, the world before or after man. From every finite thing, Proust makes a being of sensation that is constantly preserved, but by vanishing on a plane of composition of Being: "beings of flight."

EXAMPLE 13

The situation of music seems no different and perhaps embodies the frame even more powerfully. Yet it is said that sound has no frame. But compounds of sensation, sonorous blocs, equally possess sections or framing forms each of which must join together to secure a certain closing-off. The simplest cases are the melodic *air,* which is a monophonic refrain; the *motif,* which is already polyphonic, an element of a melody entering into the development of another and creating counterpoint; and the *theme,* as the object of harmonic modifications through melodic lines. These three elementary forms construct the sonorous house and its territory. They correspond to the three modalities of a being of sensation, for the air is a vibration, the motif is a clinch, a coupling, whereas the theme does not close without also unclenching, splitting, and opening. In fact, the most important musical phenomenon that appears as the sonorous compounds of sensation become more complex is that their closure or shutting-off (through the joining of their frames, of their sections) is accompanied by a possibility of opening onto an ever more limitless plane of composition. According to Bergson, musical beings are like living beings that compensate for their individuating closure by an openness created by modulation, repetition, transposition, juxtaposition. If we consider the sonata we find a particularly rigid enframing form based upon a bithematism, and in which the first movement presents the following sections: exposition of the first theme, transition, exposition of the second theme, developments on the first or second, coda, development of the first with modulation, and so on. It is an entire house with its rooms. But it is the first movement, rather, that forms a cell in this way, and great musicians rarely follow the canonical form; the other movements can open out, especially the second, through theme and variation, until Liszt ensures a fusion of movements in the "symphonic poem." The sonata appears then rather like a crossroads form where the opening of a plane of composition is born from the joining of musical sections, from the closure of sonorous compounds.

In this respect, the old procedure of theme and variation, which maintains the harmonic frame of the theme, gives way to a sort of deframing when the piano generates *compositional studies* (Chopin, Schumann, Liszt): this is a new essential moment, because creative labor no longer bears on sonorous compounds, motifs, and themes, even if this may involve extracting a plane from them, but on the contrary bears directly on the plane of composition itself, so that it gives birth to much freer and deframed compounds, to almost incomplete or overloaded aggregates, in permanent disequilibrium. Increasingly, it is the "color" of the sound that matters. We pass from the House to the Cosmos (according to a formula taken up by Stockhausen's work). The work of the plane of composition develops in two directions that involve a disaggregation of the tonal frame: the immense uniform areas [*aplats*] of continuous variation that couple and combine the forces that have become sonorous in Wagner, or the broken tones [*tons rompus*] that separate and disperse the forces by harmonizing their reversible passages in Debussy—Wagner-universe, Debussy-universe. All the tunes, all the little framing or framed refrains—childish, domestic, professional, national, territorial—are swept up in the great Refrain, a powerful song of the earth—the deterritorialized—which arises with Mahler, Berg, or Bartók. And no doubt in each case the plane of composition generates new closures, as in serial music. But, each time, the musician's action consists in deframing, in finding the opening, taking up the plane of composition once more, in accordance with the formula that obsesses Boulez: to plot a transversal, irreducible to both the harmonic vertical and melodic horizontal, that involves sonorous blocs of variable individuation but that also opens them up or splits them in a space-time that determines their density and their course over the plane.[30] The great refrain arises as we distance ourselves from the house, even if this is in order to return, since no one will recognize us any more when we come back.

Composition, composition is the sole definition of art. Composition is aesthetic, and what is not composed is not a work of art. However, technical composition, the work of the material that often calls on science (mathematics, physics, chemistry, anatomy), is not to be confused with aesthetic composition, which is the work of sensation. Only the latter fully deserves the name *composition,* and a work of art is never produced by or for the sake of technique. To be sure, technique includes many things that are individualized according to each artist and work: words and syntax in literature; not only the canvas but its preparation in painting, pigments, their mixtures, and methods of perspective; or the twelve tones of Western music, instruments, scales, and pitch. And the relationship between the two planes, between technical and aesthetic planes of composition, constantly varies historically. Take two states of oil painting that can be opposed to each other: in the first case, the picture is prepared with a white chalk background on which the outline is drawn and washed in (sketch), and finally color, light, and shade are put down. In the other case, the background becomes increasingly thick, opaque, and absorbent, so that it takes on a tinge with the wash and the work becomes impasted on a brown range, "reworkings" [*repentirs**] taking the place of the sketch: the painter paints on color, then color alongside color, increasingly the colors become accents, the architecture being assured by "the contrast of complementaries and the agreement of analogues" (van Gogh); it is through and in color that the architecture will be found, even if the accents must be given up in order to reconstitute large coloring units. It is true that Xavier de Langlais sees throughout this second case a long decline, a decadence that collapses into the ephemeral and fails to restore an architecture: the picture darkens, becomes dull, or quickly flakes.[31] And doubtless this remark raises the question, at least negatively, of progress in art, since Langlais judges decadence as beginning after Van Eyck (somewhat like those who see music coming to an end

---

*We have translated *repentirs* as "reworkings," but the French also conveys the sense of "corrections and revisions made while the painting is being executed," that is to say, not a reworking of a completed painting.

with the Gregorian chant, or philosophy with Thomas Aquinas). But it is a technical remark that concerns only the material: not only is the duration of the material quite relative but sensation belongs to a different order and possesses an existence in itself for as long as the material lasts. The relationship of sensation with the material must therefore be assessed within the limits of the duration, whatever this may be. If there is progress in art it is because art can live only by creating new percepts and affects as so many detours, returns, dividing lines, changes of level and scale. From this point of view, the distinction between two states of oil painting assumes a completely different, aesthetic and no longer technical aspect—this distinction clearly does not come down to "representational or not," since no art and no sensation have ever been representational.

In the first case *sensation is realized in the material* and does not exist outside of this realization. It could be said that sensation (the compound of sensations) is projected onto the well-prepared technical plane of composition, in such a way that the aesthetic plane of composition covers it up. The material itself must therefore include mechanisms of perspective as a result of which the projected sensation is realized not solely by covering up the picture but according to a depth. Art thus enjoys a semblance of transcendence that is expressed not in a thing to be represented but in the paradigmatic character of projection and in the "symbolic" character of perspective. According to Bergson the Figure is like fabulation: it has a religious origin. But, when it becomes aesthetic, its sensory transcendence enters into a hidden or open opposition to the suprasensory transcendence of religion.

In the second case it is no longer sensation that is realized in the material *but the material that passes into sensation.* Of course, sensation no more exists outside of this passage, and the technical plane of composition has no more autonomy, than in the first case: it is never valid for itself. But now it might be said that it *ascends* into the aesthetic plane of composition and, as Damisch says, gives it a specific thickness independent of any perspective or depth. It is at this moment that the figures of art free themselves from an apparent transcendence or paradigmatic model and avow their innocent atheism, their paganism. Of course, between these two cases, between these two states of sensation and these two poles of technique, transitions, combinations, and coexistences are constantly being produced (the impasted work of Titian or Rubens, for example): the poles are more abstract than really distinct movements. Nonetheless, modern painting, even when it is satisfied with oil and medium,* turns increasingly toward the second pole and makes the material ascend and pass "into the thickness" of the aesthetic plane of composition. That is why it is so wrong to define sensation in modern painting by the assumption of a pure visual flatness: the error is due perhaps to the fact that thickness does not need to be pronounced or deep. It could be said that Mondrian was a painter of thickness; and when Seurat defined painting as "the art of ploughing a surface," the only support he needs is the furrows and peaks of unglazed drawing paper. This is painting that no longer has any background because the "underneath" comes through: the surface can be furrowed or the plane of composition can take on thickness insofar as the material rises up, independently of depth or perspective, independently of shadows and even of the chromatic order of color (the arbitrary colorist). One no longer covers over; one raises, accumulates, piles up, goes through, stirs up, folds. It is a promotion of the ground, and sculpture can become flat since the plane is stratified. One no longer paints "on" but "under." These new powers of texture, that ascent of the ground with Dubuffet, have been pushed a long way by informal art, and by abstract expressionism and minimal art also, when they work with saturations, fibers, and layers, or when they use tarlatan or tulle in such a way that the painter can paint behind the picture in a state of

---

Médium is the same in English—"medium"—and signifies the liquid used to bind powdered color to produce paint, e.g., oil, size, egg yolk, gum arabic.

blindness.[32] With Hantaï, foldings hide from the painter's sight what, once unfolded, they give up to the spectator's eye. In any case, and in all of these states, painting is thought: vision is through thought, and the eye thinks, even more than it listens.

Hubert Damisch turned the thickness of the plane into a genuine concept by showing that "plaiting could well fulfil a role for future painting similar to that performed by perspective." This is not peculiar to painting, since Damisch finds the same distinction at the level of the architectural plane when Scarpa, for example, suppresses the movement of projection and the mechanisms of perspective so as to inscribe volumes in the thickness of the plane itself.[33] From literature to music a material thickness is affirmed that does not allow itself to be reduced to any formal depth. It is characteristic of modern literature for words and syntax to rise up into the plane of composition and hollow it out rather than carry out the operation of putting it into perspective. It is also characteristic of modern music to relinquish projection and the perspectives that impose pitch, temperament, and chromatism, so as to give the sonorous plane a singular thickness to which very diverse elements bear witness: the development of studies for the piano, which cease being just technical and become "compositional studies" (with the extension given to them by Debussy); the decisive importance assumed by the orchestra with Berlioz; the rise of timbre in Stravinsky and Boulez; the proliferation of percussive affects with metals, skins, and woods, and their combination with wind instruments to constitute blocs inseparable from the material (Varèse); the redefinition of the percept according to noise, to raw and complex sound (Cage); not only the enlargement of chromatism to other components of pitch but the tendency to a nonchromatic appearance of sound in an infinite continuum (electronic or electroacoustic music).

There is only a single plane in the sense that art includes no other plane than that of aesthetic composition: in fact, the technical plane is necessarily covered up or absorbed by the aesthetic plane of composition. It is on this condition that matter becomes expressive: either the compound of sensations is realized in the material, or the material passes into the compound, but always in such a way as to be situated on a specifically aesthetic plane of composition. There are indeed technical problems in art, and science may contribute toward their solution, but they are posed only as a function of aesthetic problems of composition that concern compounds of sensation and the plane to which they and their materials are necessarily linked. Every sensation is a question, even if the only answer is silence. In art the problem is always that of finding what monument to erect on this plane, or what plane to slide under this monument, and both at the same time: hence, in Klee, the "monument at the edge of the fertile country" and the "monument in fertile country." Are there not as many different planes as universes, authors, or even works? In fact, universes, from one art to another as much as in one and the same art, may derive from one another, or enter into relations of capture and form constellations of universes, independently of any derivation, but also scattering themselves into nebulae or different stellar systems, in accordance with qualitative distances that are no longer those of space and time. Universes are linked together or separated on their lines of flight, so that the plane may be single at the same time as universes are irreducibly multiple.

Everything (including technique) takes place between compounds of sensation and the aesthetic plane of composition. Now the latter does not come before, being neither intentional nor preconceived and having nothing to do with a program, but neither does it come afterward, although the awareness of it is formed progressively and often suddenly appears afterward. The town does not come after the house, nor the cosmos after the territory. The universe does not come after the figure, and the figure is an aptitude of a universe. We have gone from the composite sensation to the plane of composition, but only so as to recognize their strict coexistence or complementarity, neither of them advancing except through the other. The composite sensation, made up of percepts and affects, deterritorializes the system of opinion that brought together dominant perceptions and affections within a natural, historical, and social milieu. But the composite sensation is reterritorialized on the plane of composition, because it erects its houses there, because it appears there within interlocked frames or joined sections that surround its components; landscapes that have become pure percepts, and characters that become pure affects. At the same time the plane of composition involves sensation in a higher deterritorialization, making it pass through a sort of deframing which opens it up and breaks it open onto an infinite cosmos. As in Pessoa, a sensation does not occupy a place on the plane without extending it, distending it over the entire earth, and freeing all the sensations it contains: opening out or splitting open, *equaling infinity*. Perhaps the peculiarity of art is to pass through the finite in order to rediscover, to restore the infinite.

What defines thought in its three great forms—art, science, and philosophy—is always confronting chaos, laying out a plane, throwing a plane over chaos. But philosophy wants to save the infinite by giving it consistency: it lays out a plane of immanence that, through the action of conceptual personae, takes events or consistent concepts to infinity. Science, on the other hand, relinquishes the infinite in order to gain reference: it lays out a plane of simply undefined coordinates that each time, through the action of partial observers, defines states of affairs, functions, or referential propositions. Art wants to create the finite that restores the infinite: it lays out a plane of composition that, in turn, through the action of aesthetic figures, bears monuments or composite sensations. Damisch has analyzed accurately Klee's picture *Equals Infinity*. It is certainly not an allegory but the act of painting that appears as a painting. It seems to us that the brown blobs dancing in the margin and crossing the canvas are the infinite passage of chaos; the sowing of points on the canvas, divided by rods, is the finite composite sensation, but opening onto the plane of composition that restores the infinite to us, $= \infty$. However, art should not be thought to be like a synthesis of science and philosophy, of the finite and infinite routes. The three routes are specific, each as direct as the others, and they are distinguished by the nature of the plane and by what occupies it. Thinking is thought through concepts, or functions, or sensations and no one of these thoughts is better than another, or more fully, completely, or synthetically "thought." The frames of art are no more scientific coordinates than sensations are concepts, or vice versa. Abstract art and conceptual art are two recent attempts to bring art and philosophy together, but they do not substitute the concept for the sensation; rather they create sensations and not concepts. Abstract art seeks only to refine sensation, to dematerialize it by setting out an architectonic plane of composition in which it would become a purely spiritual being, a radiant thinking and thought matter, no longer a sensation of sea or tree, but a sensation of the concept of sea or concept of tree. Conceptual art seeks an opposite dematerialization through generalization, by installing a sufficiently neutralized plane of composition (the catalog that brings together works not displayed, the ground covered by its own map, disused spaces without architecture and the "flatbed"* plane) so that everything takes on a value of sensation reproducible to infinity: things, images or clichés, propositions—a thing, its photograph on the same scale and in the same place, its dictionary definition. However, in the latter case it is not at all clear that this way leads either to the sensation or to the concept, because the plan of composition tends to become "informative," and the sensation depends upon the simple "opinion" of a spectator who determines whether or not to "materialize" the sensation, that is to say, decides whether or not it is art. This is a lot of effort to find ordinary perceptions and affections in the infinite and to reduce the concept to a *doxa* of the social body or great American metropolis.

The three thoughts intersect and intertwine but without synthesis

---

*In English in the original.

or identification. With its concepts, philosophy brings forth events. Art erects monuments with its sensations. Science constructs states of affairs with its functions. A rich tissue of correspondences can be established between the planes. But the network has its culminating points, where sensation itself becomes sensation of concept or function, where the concept becomes concept of function or of sensation, and where the function becomes function of sensation or concept. And none of these elements can appear without the other being still to come, still indeterminate or unknown. Each created element on a plane calls on other heterogeneous elements, which are still to be created on other planes: thought as heterogenesis. It is true that these culminating points contain two extreme dangers: either leading us back to the opinion from which we wanted to escape or precipitating us into the chaos that we wanted to confront.

We require just a little order to protect us from
chaos. Nothing is more distressing than a
thought that escapes itself, than ideas that fly off,
that disappear hardly formed, already eroded by
forgetfulness or precipitated into others that we
no longer master. These are infinite *variabilities,*
the appearing and disappearing of which coincide. They are infinite speeds that blend into the
immobility of the colorless and silent nothingness
they traverse, without nature or thought. This is
the instant of which we do not know whether it
is too long or too short for time. We receive sudden jolts that beat like arteries. We constantly
lose our ideas. That is why we want to hang on
to fixed opinions so much. We ask only that our
ideas are linked together according to a minimum of constant rules. All that the association of
ideas has ever meant is providing us with these
protective rules—resemblance, contiguity, causality—which enable us to put some order into
ideas, preventing our "fantasy" (delirium, madness) from crossing the universe in an instant,
producing winged horses and dragons breathing

fire. But there would not be a little order in ideas if there was not also a little order in things or states of affairs, like an objective antichaos: "If cinnabar were sometimes red, sometimes black, sometimes light, sometimes heavy . . . , my empirical imagination would never find opportunity when representing red color to bring to mind heavy cinnabar."[1] And finally, at the meeting point of things and thought, the sensation must recur—that of heaviness whenever we hold cinnabar in our hands, that of red whenever we look at it—as proof or evidence of their agreement with our bodily organs that do not perceive the present without imposing on it a conformity with the past. This is all that we ask for in order to *make an opinion* for ourselves, like a sort of "umbrella," which protects us from chaos.

Our opinions are made up from all this. But art, science, and philosophy require more: they cast planes over the chaos. These three disciplines are not like religions that invoke dynasties of gods, or the epiphany of a single god, in order to paint a firmament on the umbrella, like the figures of an *Urdoxa* from which opinions stem. Philosophy, science, and art want us to tear open the firmament and plunge into the chaos. We defeat it only at this price. And thrice victorious I have crossed the Acheron. The philosopher, the scientist, and the artist seem to return from the land of the dead. What the philosopher brings back from the chaos are *variations* that are still infinite but that have become inseparable on the absolute surfaces or in the absolute volumes that lay out a secant [*sécant*] plane of immanence: these are not associations of distinct ideas, but reconnections through a zone of indistinction in a concept. The scientist brings back from the chaos *variables* that have become independent by slowing down, that is to say, by the elimination of whatever other variabilities are liable to interfere, so that the variables that are retained enter into determinable relations in a function: they are no longer links of properties in things, but finite coordinates on a secant plane of reference that go from local probabilities to a global cosmology. The artist brings back from the chaos *varieties* that no longer constitute a reproduction of the sensory in the organ but set up a being of the sensory, a being of sensation, on an anorganic plane of composition that is able to restore the infinite. The struggle with chaos that Cézanne and Klee have shown in action in painting, at the heart of painting, is found in another way in science and in philosophy: it is always a matter of defeating chaos by a secant plane that crosses it. Painters go through a catastrophe, or through a conflagration, and leave the trace of this passage on the canvas, as of the leap that leads them from chaos to composition.[2] Mathematical equations do not enjoy a tranquil certainty, which would be like the sanction of a dominant scientific opinion, but arise from an abyss that makes the mathematician "readily skip over calculations," in anticipation of not being able to bring about or arrive at the truth without "colliding here and there."[3] And philosophical thought does not bring its concepts together in friendship without again being traversed by a fissure that leads them back to hatred or disperses them in the coexisting chaos where it is necessary to take them up again, to seek them out, to make a leap. It is as if one were casting a net, but the fisherman always risks being swept away and finding himself in the open sea when he thought he had reached port. The three disciplines advance by crises or shocks in different ways, and in each case it is their succession that makes it possible to speak of "progress." It is as if the *struggle against chaos* does not take place without an affinity with the enemy, because another struggle develops and takes on more importance—the struggle *against opinion,* which claims to protect us from chaos itself.

In a violently poetic text, Lawrence describes what produces poetry: people are constantly putting up an umbrella that shelters them and on the underside of which they draw a firmament and write their conventions and opinions. But poets, artists, make a slit in the umbrella, they tear open the firmament itself, to let in a bit of free and windy chaos and to frame in a sudden light a vision that appears through the rent—Wordsworth's spring or Cézanne's apple, the silhouettes of Macbeth or Ahab. Then come the crowd of imitators who repair the umbrella with something vaguely resembling the vision, and the crowd of commentators who patch over the rent with opinions: communication. Other artists are always needed to make other slits, to carry out necessary and perhaps ever-greater destructions, thereby restoring to their predecessors the incommunicable novelty that we could no longer see. This is to say that artists struggle less against chaos (that, in a certain manner, all their wishes summon forth) than against the "clichés" of opinion.[4] The painter does not paint on an empty canvas, and neither does the writer write on a blank page; but the page or canvas is already so covered with preexisting, preestablished clichés that it is first necessary to erase, to clean, to flatten, even to shred, so as to let in a breath of air from the chaos that brings us the vision. When Fontana slashes the colored canvas with a razor, he does not tear the color in doing this. On the contrary, he makes us see the area of plain, uniform color, of pure color, through the slit. Art indeed struggles with chaos, but it does so in order to bring forth a vision that illuminates it for an instant, a Sensation. Even houses: Soutine's drunken houses come from chaos, knocking up against one another and preventing one another from falling back into it; Monet's house also rises up like a slit through which chaos becomes the vision of roses. Even the most delicate pink opens on to chaos, like flesh on the flayed body.[5] A work of chaos is certainly no better than a work of opinion; art is no more made of chaos than it is of opinion. But if art battles against chaos it is to borrow weapons from it that it turns against opinion, the better to defeat it with tried and tested arms. Because the picture starts out covered with clichés, the painter must confront the chaos and hasten the destructions so as to produce a sensation that defies every opinion and cliché (how many times?). Art is not chaos but a composition of chaos that yields the vision or sensation, so that it constitutes, as Joyce says, a chaosmos, a composed chaos—neither foreseen nor preconceived. Art transforms chaotic variability into *chaoid* variety, as in El Greco's black and green-gray conflagration, for example, or Turner's golden conflagration, or de Staël's red conflagration. Art struggles with chaos but it does so in order to render it sensory, even through the most charming character, the most enchanted landscape (Watteau).

Science is perhaps inspired by a similar sinuous, reptilian movement. A struggle against chaos seems to be an essential part of science when it puts slow variability under constants or limits, when it thereby refers it to centers of equilibrium, when it subjects it to a selection that retains only a small number of independent variables within coordinate axes, and when between these variables it installs relationships whose future state can be determined on the basis of the present (determinist calculus) or, alternatively, when it introduces so many variables at once that the state of affairs is only statistical (calculus of probabilities). In this sense we speak of a specifically scientific opinion won from chaos, as we do of a communication defined sometimes by initial pieces of information, sometimes by large-scale pieces of information, which usually go from the elementary to the composite, or from the present to the future, or from the molecular to the molar. But, here again, science cannot avoid experiencing a profound attraction for the chaos with which it battles. If slowing down is the thin border that separates us from the oceanic chaos, science draws as close as it can to the nearest waves by positing relationships that are preserved with the appearance and disappearance of variables (differential calculus). The difference between the chaotic state where the appearance and disappearance of a variability blend together, and the semichaotic state that manifests a relationship as the limit of the variables that appear or disappear becomes ever smaller. As Michel Serres says of Leibniz, "There would be two infraconsciousnesses: the deeper would be structured like any set whatever, a pure multiplicity or possibility in general, an aleatory mixture of signs; the less deep would be covered by combinatory schemas of this multiplicity."[6] One could conceive of a series of coordinates or phase spaces as a succession of filters, the earlier of which would be in each case a relatively chaotic state, and the later a chaoid state, so that we would cross chaotic thresholds rather than go from the elementary to the composite. Opinion offers us a science that dreams of unity, of unifying its laws, and that still searches today for a community of the four forces. Nevertheless, the dream of capturing a bit of chaos is more insistent, even if the most diverse forces stir restlessly within it. Science would relinquish all the rational unity to which it aspires for a little piece of chaos that it could explore.

Art takes a bit of chaos in a frame in order to form a composed chaos that becomes sensory, or from which it extracts a chaoid sensation as variety; but science takes a bit of chaos in a system of coordinates and forms a referenced chaos that becomes Nature, and from which it extracts an aleatory function and chaoid variables. In this way one of the most important aspects of modern mathematical physics appears in the action of "strange" or chaotic attractors: two neighboring trajectories in a determinate system of coordinates do not remain so and diverge in an exponential manner before coming together through operations of stretching and folding that are repeated and intersect with chaos.[7] If equilibrium attractors (fixed points, limit cycles, cores) express science's struggle with chaos, strange attractors reveal its profound attraction to chaos, as well as the constitution of a chaosmos internal to modern science (everything that, in one way or another, was misrepresented in earlier periods, notably in the fascination for turbulences). We thus come back to a conclusion to which art led us: the struggle with chaos is only the instrument of a more profound struggle against opinion, for the misfortune of people comes from opinion. Science turns against opinion, which lends to it a religious taste for unity or unification. But it also turns within itself against properly scientific opinion as *Urdoxa,* which consists sometimes in determinist prediction (Laplace's God) and sometimes in probabilistic evaluation (Maxwell's demon): by releasing itself from initial pieces of information and large-scale pieces of information, science substitutes for communication the conditions of creativity defined by singular effects and minimal fluctuations. Creation is the aesthetic varieties or scientific variables that emerge on a plane that is able to crosscut chaotic variability. As for pseudosciences that claim to study the phenomena of opinion, the artificial intelligences of which they make use maintain as their models probabilistic processes, stable attractors, an entire logic of the recognition of forms; but they must achieve chaoid states and chaotic attractors to be able to understand both thought's struggle against opinion and its degeneration into opinion (one line in the development of computers is toward the assumption of a chaotic or chaoticizing system).

This is what confirms the third case, which is no longer sensory variety or functional variable but conceptual variation as it appears in philosophy. Philosophy struggles in turn with the chaos as undifferentiated abyss or ocean of dissemblance. But this does not mean that philosophy ranges itself on the side of opinion, nor that opinion can take its place. A concept is not a set of associated ideas like an opinion. Neither is it an order of reasons, a series of ordered reasons that could rigorously constitute a kind of rationalized *Urdoxa*. To reach the concept it is not even enough for phenomena to be subject to principles analogous to those that associate ideas or things, or to principles that order reasons. As Michaux says, what suffices for "current ideas" does not suffice for "vital ideas"—those that must be created. Ideas can only be associated as images and can only be ordered as abstractions; to arrive at the concept we must go beyond both of these and arrive *as quickly as possible* at mental objects determinable as real beings. This is what Spinoza or Fichte have already shown: we must make use of fictions and abstractions, but only so far as is necessary to get to a plane where we go from real being to real being and advance through the construction of concepts.[8] We have seen how this result can be achieved to the extent that variations become inseparable according to zones of neighborhood or indiscernibility: they then cease being associable according

to the caprice of imagination, or discernible and capable of being
ordered according to the exigencies of reason, in order to form genuine conceptual blocs. A concept is a set of inseparable variations that
is produced or constructed on a plane of immanence insofar as the
latter crosscuts the chaotic variability and gives it consistency (reality). A concept is therefore a chaoid state par excellence; it refers back
to a chaos rendered consistent, become Thought, mental chaosmos.
And what would *thinking* be if it did not constantly confront chaos?
Reason shows us its true face only when it "thunders in its crater."
Even the cogito is only an opinion, an *Urdoxa* at best, if we do not
extract from it the inseparable variations that make it a concept, if we
do not give up finding an umbrella or shelter in it, unless we stop
presupposing an immanence that would be accommodated *to itself*,
so that, on the contrary, it can set itself up on a plane of immanence
to which it belongs that which takes it back to the open sea. In short,
chaos has three daughters, depending on the plane that cuts through
it: these are the *Chaoids*—art, science, and philosophy—as forms of
thought or creation. We call *Chaoids* the realities produced on the
planes that cut through the chaos in different ways.

The brain is the junction—not the unity—*of the three planes.* Certainly, when the brain is considered as a determinate function it
appears as a complex set both of horizontal connections and of vertical
integrations reacting on one another, as is shown by cerebral "maps."
The question, then, is a double one: are the connections preestablished, as if guided by rails, or are they produced and broken up in
fields of forces? And are the processes of integration localized hierarchical centers, or are they rather forms (*Gestalten*) that achieve their
conditions of stability in a field on which the position of center itself
depends? In this respect the importance of Gestalt theory concerns
the theory of the brain just as much as the conception of perception,
since it is directly opposed to the status of the cortex as it appears
from the point of view of conditioned reflexes. But, whatever point of
view is considered, it is not difficult to show that similar difficulties

are encountered whether paths are ready-made or self-producing, and whether centers are mechanical or dynamical. Ready-made paths that are followed step by step imply a preestablished track, but trajectories constituted within a field of forces proceed through resolution of tensions also acting step by step (for example, the tension of reconciliation between the fovea and the luminous point projected on the retina, the latter having a structure analogous to a cortical area): both schemas presuppose a "plane," not an end or a program, but a *survey of the entire field*. This is what Gestalt theory does not explain, any more than mechanism explains preassembly [*prémontage*].

It is not surprising that the brain, treated as a constituted object of science, can be an organ only of the formation and communication of opinion: this is because step-by-step connections and centered integrations are still based on the limited model of recognition (gnosis and praxis; "this is a cube"; "this is a pencil"), and the biology of the brain is here aligned on the same postulates as the most stubborn logic. Opinions are pregnant forms, like soap bubbles according to the Gestalt, with regard to milieus, interests, beliefs, and obstacles. Thus it seems difficult to treat philosophy, art, and even science as "mental objects," simple assemblages of neurones in the objectified brain, since the derisory model of recognition confines these latter within the *doxa*. If the mental objects of philosophy, art, and science (that is to say, vital ideas) have a place, it will be in the deepest of the synaptic fissures, in the hiatuses, intervals, and meantimes of a nonobjectifiable brain, in a place where to go in search of them will be to create. It will be a bit like tuning a television screen whose intensities would bring out that which escapes the power of objective definition.[9] That is to say, thought, even in the form it actively assumes in science, does not depend upon a brain made up of organic connections and integrations: according to phenomenology, thought depends on man's relations with the world—with which the brain is necessarily in agreement because it is drawn from these relations, as excitations are drawn from the world and reactions from man, including their uncertainties and failures. "Man thinks, not the brain"; but this ascent of phenomenology beyond the brain toward a Being in the world, through a double criticism of mechanism and dynamism, hardly gets us out of the sphere of opinions. It leads us only to an *Urdoxa* posited as original opinion, or meaning of meanings.[10]

Will the turning point not be elsewhere, in the place where the brain is "subject," where it becomes subject? It is the brain that thinks and not man—the latter being only a cerebral crystallization. We will speak of the brain as Cézanne spoke of the landscape: man absent from, but completely within the brain. Philosophy, art, and science are not the mental objects of an objectified brain but the three aspects under which the brain becomes subject, Thought-brain. They are the three planes, the rafts on which the brain plunges into and confronts the chaos. What are the characteristics of this brain, which is no longer defined by connections and secondary integrations? It is not a brain behind the brain but, first of all, a state of survey without distance, at ground level, a self-survey that no chasm, fold, or hiatus escapes. It is a primary, "true form" as Ruyer defined it: neither a Gestalt nor a perceived form but a *form in itself* that does not refer to any external point of view, any more than the retina or striated area of the cortex refers to another retina or cortical area; it is an absolute consistent form that surveys *itself* independently of any supplementary dimension, which does not appeal therefore to any transcendence, which has only a single side whatever the number of its dimensions, which remains copresent to all its determinations without proximity or distance, traverses them at infinite speed, without limit-speed, and which makes of them so many *inseparable variations* on which it confers an equipotentiality without confusion.[11] We have seen that this was the status of the concept as pure event or reality of the virtual. And doubtless concepts are not limited to just one and the same brain since each one of them constitutes a "domain of survey," and the transitions from one concept to another remain irreducible insofar as a new concept does not render its copresence or equipotentiality of determinations necessary in turn. Nor will we say that every concept is a brain. But the brain, under its first aspect of absolute form, appears as the faculty of concepts, that is to say, as the faculty of their creation, at the same time that it sets up the plane of immanence on which concepts are placed, move, change order and relations, are renewed, and never cease being created. The brain is the *mind* itself. At the same time that the brain becomes subject—or rather "superject," as Whitehead puts it—the concept becomes object as created, as event or creation itself; and philosophy becomes the plane of immanence that supports the concepts and that the brain lays out. Cerebral movements also give rise to conceptual personae.

It is the brain that says *I*, but *I* is an other. It is not the same brain as the brain of connections and secondary integrations, although there is no transcendence here. And this *I* is not only the "I conceive" of the brain as philosophy, it is also the "I feel" of the brain as art. Sensation is no less brain than the concept. If we consider the nervous connections of excitation-reaction and the integrations of perception-action, we need not ask at what stage on the path or at what level sensation appears, for it is presupposed and withdrawn. The withdrawal is not the opposite but a correlate of the survey. Sensation is excitation itself, not insofar as it is gradually prolonged and passes into the reaction but insofar as it is preserved or preserves its vibrations. Sensation contracts the vibrations of the stimulant on a nervous surface or in a cerebral volume: what comes before has not yet disappeared when what follows appears. This is its way of responding to chaos. Sensation itself vibrates because it contracts vibrations. It preserves itself because it preserves vibrations: it is Monument. It resonates because it makes its harmonics resonate. Sensation is the contracted vibration that has become quality, variety. That is why the brain-subject is here called *soul* or *force*, since only the soul preserves by contracting that which matter dissipates, or radiates, furthers, reflects, refracts, or converts. Thus the search for sensation is fruitless if we go no farther than reactions and the excitations that they prolong, than actions and the perceptions that they reflect: this is because the soul (or rather, the force), as Leibniz said, does nothing, or does not act, but is only present; it preserves. Contraction is not an action but a pure passion, a contemplation that preserves the before in the after.[12] Sensation, then, is on a plane that is different from mechanisms, dynamisms, and finalities: it is on a plane of composition where sensation is formed by contracting that which composes it, and by composing itself with other sensations that contract it in turn. Sensation is pure contemplation, for it is through contemplation that one contracts, contemplating oneself to the extent that one contemplates the elements from which one originates. Contemplating is creating, the mystery of passive creation, sensation. Sensation fills out the plane of composition and is filled with itself by filling itself with what it contemplates: it is "enjoyment" and "self-enjoyment."* It is a subject, or rather an *inject*. Plotinus defined all things as contemplations, not only people and animals but plants, the earth, and rocks. These are not Ideas that we contemplate through concepts but the elements of matter that we contemplate through sensation. The plant contemplates by contracting the elements from which it originates—light, carbon, and the salts—and it fills itself with colors and odors that in each case qualify its variety, its composition: it is sensation in itself.[13] It is as if flowers smell themselves by smelling what composes them, first attempts of vision or of sense of smell, before being perceived or even smelled by an agent with a nervous system and a brain.

Of course, plants and rocks do not possess a nervous system. But, if nerve connections and cerebral integrations presuppose a brain-force as faculty of feeling coexistent with the tissues, it is reasonable to suppose also a faculty of feeling that coexists with embryonic tissues and that appears in the Species as a collective brain; or with the vegetal tissues in the "small species." Chemical affinities and

*In English in the original.

physical causalities themselves refer to primary forces capable of preserving their long chains by contracting their elements and by making them resonate: no causality is intelligible without this subjective
instance. Not every organism has a brain, and not all life is organic,
but everywhere there are forces that constitute microbrains, or an
inorganic life of things. We can dispense with Fechner's or Conan
Doyle's splendid hypothesis of a nervous system of the earth only
because the force of contracting or of preserving, that is to say, of
feeling appears only as a global brain in relation to the elements
contracted directly and to the mode of contraction, which differ depending on the domain and constitute precisely irreducible varieties.
But, in the final analysis, the same ultimate elements and the same
withdrawn force constitute a single plane of composition bearing all
the varieties of the universe. Vitalism has always had two possible
interpretations: that of an Idea that acts, but is not—that acts therefore only from the point of view of an external cerebral knowledge
(from Kant to Claude Bernard); or that of a force that is but does not
act—that is therefore a pure internal Awareness (from Leibniz to
Ruyer). If the second interpretation seems to us to be imperative it is
because the contraction that preserves is always in a state of detach-
ment in relation to action or even to movement and appears as a pure
contemplation without knowledge. This can be seen even in the
cerebral domain par excellence of apprenticeship or the formation of
habits: although everything seems to take place by active connections
and progressive integrations, from one test to another, the tests or
cases, the occurrences, must, as Hume showed, be contracted in a
contemplating "imagination" while remaining distinct in relation to
actions and to knowledge. Even when one is a rat, it is through
contemplation that one "contracts" a habit. It is still necessary to
discover, beneath the noise of actions, those internal creative sensations or those silent contemplations that bear witness to a brain.

These first two aspects or layers of the brain-subject, sensation as
much as the concept, are very fragile. Not only objective disconnec-

tions and disintegrations but an immense weariness results in sensations, which have now become woolly, letting escape the elements and vibrations it finds increasingly difficult to contract. Old age is this very weariness: then, there is either a fall into mental chaos outside of the plane of composition or a falling-back on ready-made opinions, on cliches that reveal that an artist, no longer able to create new sensations, no longer knowing how to preserve, contemplate, and contract, no longer has anything to say. The case of philosophy is a bit different, although it depends upon a similar weariness. In this case, weary thought, incapable of maintaining itself on the plane of immanence, can no longer bear the infinite speeds of the third kind that, in the manner of a vortex, measure the concept's copresence to all its intensive components at once (consistency). It falls back on the relative speeds that concern only the succession of movement from one point to another, from one extensive component to another, from one idea to another, and that measure simple associations without being able to reconstitute any concept. No doubt these relative speeds may be very great, to the point of simulating the absolute, but they are only the variable speeds of opinion, of discussion or "repartee," as with those untiring young people whose mental quickness is praised, but also with those weary old ones who pursue slow-moving opinions and engage in stagnant discussions by speaking all alone, within their hollowed head, like a distant memory of their old concepts to which they remain attached so as not to fall back completely into the chaos.

No doubt, as Hume says, causalities, associations, and integrations inspire opinions and beliefs in us that are ways of expecting and recognizing something (including "mental objects"): it will rain, the water will boil, this is the shortest route, this is the same figure from a different view. But, although such opinions frequently slip in among scientific propositions, they do not form part of them; and science subjects these processes to operations of a different nature, which constitute an activity of knowing and refer to a faculty of knowledge as the third layer of a brain-subject that is no less creative than the other two. Knowledge is neither a form nor a force but a *function:* "I function." The subject now appears as an "eject," because it extracts elements whose principal characteristic is distinction, discrimination: limits, constants, variables, and functions, all those functives and prospects that form the terms of the scientific proposition. Geometrical projections, algebraic substitutions and transformations consist not in recognizing something through variations but in distinguishing variables and constants, or in progressively discriminating the terms that tend toward successive limits. Hence, when a constant is assigned in a scientific operation, it is not a matter of contracting cases or moments in a single contemplation but one of establishing a necessary relation between factors that remain independent. The fundamental actions of the scientific faculty of knowledge appear to us in this sense to be the following: setting limits that mark a renunciation of infinite speeds and lay out a plane of reference; assigning variables that are organized in series tending toward these limits; coordinating the independent variables in such a way as to establish between them or their limits necessary relations on which distinct functions depend, the plane of reference being a coordination in actuality; determining mixtures or states of affairs that are related to the coordinates and to which functions refer. It is not enough to say that these operations of scientific knowledge are functions of the brain; the functions are themselves the folds of a brain that lay out the variable coordinates of a plane of knowledge (reference) and that dispatch partial observers everywhere.

There is still an operation that clearly shows the persistence of chaos, not only around the plane of reference or coordination but in the detours of its variable surface, which are always put back into play. These are operations of branching and individuation: if states of affairs are subject to them it is because they are inseparable from the potentials they take from chaos itself and that they do not actualize without risk of dislocation or submergence. It is therefore up to science to make evident the chaos into which the brain itself, as subject of knowledge, plunges. The brain does not cease to constitute limits that determine functions of variables in particularly extended areas; relations between these variables (connections) manifest all the more an uncertain and hazardous characteristic, not only in electrical synapses, which show a statistical chaos, but in chemical synapses, which refer to a deterministic chaos.[14] There are not so much cerebral centers as points, concentrated in one area and disseminated in another, and "oscillators," oscillating molecules that pass from one point to another. Even in a linear model like that of the conditioned reflex, Erwin Straus has shown that it was essential to understand the intermediaries, the hiatuses and gaps. Arborized paradigms give way to rhizomatic figures, acentered systems, networks of finite automatons, chaoid states. No doubt this chaos is hidden by the reinforcement of opinion generating facilitating paths, through the action of habits or models of recognition; but it will become much more noticeable if, on the contrary, we consider creative processes and the bifurcations they imply. And individuation, in the cerebral state of affairs, is all the more functional because it does not have the cells themselves for variables, since the latter constantly die without being renewed, making the brain a set of little deaths that puts constant death within us. It calls upon a potential that is no doubt actualized in the determinable links that derive from perceptions, but even more in the free effect that varies according to the creation of concepts, sensations, or functions themselves.

The three planes, along with their elements, are irreducible: *plane of immanence of philosophy, plane of composition of art, plane of reference or coordination of science; form of concept, force of sensation, function of knowledge; concepts and conceptual personae, sensations and aesthetic figures, figures and partial observers.* Analogous problems are posed for each plane: in what sense and how is the plane, in each case, one or multiple—what unity, what multiplicity? But what to us seem more important now are the problems of interference between the planes that join up in the brain. A first type of interference appears when a philosopher attempts to create the concept of a sensation or a function (for example, a concept peculiar to Riemannian space or to irrational number); or when a scientist tries to create functions of sensations, like Fechner or in theories of color or sound, and even functions of concepts, as Lautman demonstrates for mathematics insofar as the latter actualizes virtual concepts; or when an artist creates pure sensations of concepts or functions, as we see in the varieties of abstract art or in Klee. In all these cases the rule is that the interfering discipline must proceed with its own methods. For example, sometimes we speak of the intrinsic beauty of a geometrical figure, an operation, or a demonstration, but so long as this beauty is defined by criteria taken from science, like proportion, symmetry, dissymmetry, projection, or transformation, then there is nothing aesthetic about it: this what Kant demonstrated with such force.[15] The function must be grasped within a sensation that gives it percepts and affects composed exclusively by art, on a specific plane of creation that wrests it from any reference (the intersection of two black lines or the thickness of color in the right angles in Mondrian; or the approach of chaos through the sensation of strange attractors in Noland or Shirley Jaffe).

These, then, are extrinsic interferences, because each discipline remains on its own plane and utilizes its own elements. But there is a second, intrinsic type of interference when concepts and conceptual personae seem to leave a plane of immanence that would correspond to them, so as to slip in among the functions and partial observers, or among the sensations and aesthetic figures, on another plane; and similarly in the other cases. These slidings are so subtle, like those of Zarathustra in Nietzsche's philosophy or of Igitur in Mallarmé's poetry, that we find ourselves on complex planes that are difficult to qualify. In turn, partial observers introduce into science sensibilia that are sometimes close to aesthetic figures on a mixed plane.

Finally, there are interferences that cannot be localized. This is because each distinct discipline is, in its own way, in relation with a negative: even science has a relation with a nonscience that echoes its effects. It is not just a question of saying that art must form those of us who are not artists, that it must awaken us and teach us to feel, and that philosophy must teach us to conceive, or that science must teach us to know. Such pedagogies are only possible if each of the disciplines is, on its own behalf, in an essential relationship with the No that concerns it. The plane of philosophy is prephilosophical insofar as we consider it in itself independently of the concepts that come to occupy it, but nonphilosophy is found where the plane confronts chaos. *Philosophy needs a nonphilosophy that comprehends it; it needs a nonphilosophical comprehension just as art needs nonart and science needs nonscience.*[16] They do not need the No as beginning, or as the end in which they would be called upon to disappear by being realized, but at every moment of their becoming or their development. Now, if the three Nos are still distinct in relation to the cerebral plane, they are no longer distinct in relation to the chaos into which the brain plunges. In this submersion it seems that there is extracted from chaos the shadow of the "people to come" in the form that art, but also philosophy and science, summon forth: mass-people, world-people, brain-people, chaos-people—nonthinking thought that lodges in the three, like Klee's nonconceptual concept or Kandinsky's internal silence. It is here that concepts, sensations, and functions become undecidable, at the same time as philosophy, art, and science become indiscernible, as if they shared the same shadow that extends itself across their different nature and constantly accompanies them.

Notes

Introduction: The Question Then . . .

1. See *L'oeuvre ultime de Cézanne à Dubuffet* (Saint-Paul-de-Vence: Fondation Maeght, 1976), with preface by Jean-Louis Prat.

2. Pierre Barbéris, *Chateaubriand: Une réaction au monde moderne* (Paris: Larousse, 1976): "*Rancé,* a book on old age as impossible value, is a book written against old age in power; it is a book of universal ruins in which only the power of writing is affirmed."

3. Alexandre Kojève, "Tyranny and Wisdom," in Leo Strauss, *On Tyranny,* p. 156 (New York: Cornell University Press, 1968).

4. For example, Xenophon, *Constitution of the Lacedaemonians,* 4.5. These aspects of the city have been analyzed by Detienne and Vernant.

5. On the relationship of friendship with the possibility of thought in the modern world, see Maurice Blanchot, *L'amitié* (Paris: Gallimard, 1971), and the dialogue between two weary men in Maurice Blanchot, *The Infinite Conversation,* trans. Susan Hanson (Minneapolis: University of Minnesota Press, 1993). See also Dionys Mascolo, *Autour d'un effort de mémoire* (Paris: Nadeau, 1987).

6. F. Nietzsche, *The Will to Power,* trans. Walter Kaufman and R. J. Hollingdale (New York: Vintage, 1968), 409.

7. Plato, *The Statesman,* 268a, 279a.

8. In a form that is deliberately like a schoolbook, Frédéric Cossutta has proposed a very interesting pedagogy of the concept: Frédéric Cossutta, *Eléments pour la lecture des textes philosophiques* (Paris: Bordas, 1989).

1: What Is a Concept?

1. This history, which does not begin with Leibniz, passes through episodes as diverse as the constant

theme of the proposition of the other person in Wittgenstein ("he has tooth-ache . . .") and the position of the other person as theory of possible world in Michel Tournier, *Friday, or The Other Island* (Harmondsworth: Penguin, 1974).

2. On the survey and absolute surfaces or volumes as real beings, cf. Raymond Ruyer, *Néo-finalisme* (Paris: P.U.F., 1952) chaps. 9–11.

3. G. W. Leibniz, "New System and Explanation of the New System," in *Philosophical Writings,* ed. G.H.R. Parkinson (London: Everyman's Library, 1973), p. 121.

4. Gilles-Gaston Granger, *Pour la connaisance philosophique* (Paris: Odile Jacob, 1988), chap. 6.

2: The Plane of Immanence

1. On the elasticity of the concept, see Hubert Damisch, preface to Jean Dubuffet, *Prospectus et tous écrits suivants* (Paris: Gallimard, 1967), vol. 1, pp. 18–19.

2. Jean-Pierre Luminet distinguishes relative horizons, like the terrestrial horizon centered on, and changing with, an observer, and the absolute horizon, the "horizon of events," which is independent of any observer and distributes events into two categories, seen and nonseen, communicable and noncommunicable. "Le trou noir et l'infini," in *Les dimensions de l'infini* (Paris: Institut culturel italien de Paris, n.d.). We refer also to the Zen text of the Japanese monk Dôgen, which invokes the horizon or "reserve" of events: *Shôbogenzo,* trans. and with commentary by René de Ceccaty and Ryôji Nakamura (Paris: La Différence, 1980).

3. Epicurus, "Letter to Herodotus," 61–62, in *Letters, Principal Doctrines, and Vatican Sayings,* trans. Russel M. Geer, (Indianapolis: Bobbs-Merrill, 1964), p. 22

4. On these dynamisms see Michel Courthial's forthcoming *Le visage.*

5. François Laruelle is engaged in one of the most interesting undertakings of contemporary philosophy. He invokes a One-All that he qualifies as "nonphilosophical" and, oddly, as "scientific," on which the "philosophical decision" takes root. This One-All seems to be close to Spinoza. François Laruelle, *Philosophie et non-philosophie* (Liege: Mardaga, 1989).

6. In 1939 Etienne Souriau published *L'instauration philosophique* (Paris: Alcan, 1939). Aware of creative activity in philosophy, he invoked a kind of plane of instituting as the ground of this creation, or "philosopheme," animated by dynamisms (pp. 62–63).

7. Cf. Jean-Pierre Vernant, *The Origins of Greek Thought* (Ithaca: Cornell University Press, 1982), pp. 107–29.

8. Immanuel Kant, *Critique of Pure Reason,* trans. N. Kemp-Smith (Lon-

don: Macmillan, 1929): space as form of exteriority is no less "in ourselves" than time as form of interiority ("Critique of the Fourth Paralogism of Transcendental Psychology"). On the Idea as "horizon," cf. "Appendix to the Transcendental Dialectic."

9. Raymond Bellour, *L'entre-images: photo, cinéma, vidéo* (Paris: La Différence, 1990), p. 132, on the link between transcendence and the interruption of movement or the freeze-frame [*arrêt sur image*].

10. Jean-Paul Sartre, *The Transcendence of the Ego,* trans. Forrest Williams and Robert Kirkpatrick (New York: Noonday Press, 1957), p. 23 (reference to Spinoza).

11. Antonin Artaud, *The Peyote Dance* (a translation of *Les Tarahumaras*), trans. Helen Weaver (New York: Farrar, Straus and Giroux, 1976).

12. E. Naville, *Maine De Biran, sa vie et ses pensées* (Paris: Naville, 1857), p. 357.

13. Cf. Heinrich von Kleist, "De l'élaboration progressive des idées dans le discours," in *Anecdotes et petits écrits* (Paris: Payot, 1981), p. 77; and Artaud, "Correspondence with Rivière," in Antonin Artaud, *Collected Works,* trans. V. Corti (London: Calder and Boyars, 1968), vol. 1.

14. *Jean Tinguely: Swiss Sculptures* (Paris: Centre Georges Pompidou, Musée nationale d'art moderne, 1988).

15. Maurice Blanchot, *The Infinite Conversation,* trans. Susan Hanson (Minneapolis: University of Minnesota Press, 1993), p. 46. On the unthought in thought, see Michel Foucault, *The Order of Things* (London: Tavistock, 1970), pp. 322–28. See also the "distant interior" in Michaux.

3: Conceptual Personae

1. On the Idiot (the uninitiated, private, or ordinary individual as opposed to the technician or expert) in his relationships with thought, see Nicholas of Cusa, *The Idiot* (trans. of *Idiota* [1450]; London, 1650). Descartes reconstitutes the three personae under the names of Eudoxus, the idiot; Polyander, the technician; and Epistemon, the public expert. "The Search for Truth by Means of the Natural Light," in *The Philosophical Writings of Descartes,* trans. John Cottingham, Robert Stoothoff, and Dugald Murdoch (Cambridge: Cambridge University Press, 1984), vol. 2.

2. Leon Chestov takes the new opposition from Kierkegaard, first of all: *Kierkegaard et la philosophie existentielle,* trans. T. Rageot and B. de Schloezer (Paris: Vrin, 1972).

3. Herman Melville, *The Confidence-Man: His Masquerade* (Oxford: Oxford University Press, 1989), chap. 44.

4. Michel Guérin, *La terreur et la pitié* (Arles: Actes Sud, 1990).

5. Cf. the analyses of Isaac Joseph, who draws on Simmel and Goffman: *Le passant considérable: essai sur la dispersion de l'espace public* (Paris: Méridiens, 1984).

6. On the persona of the Stranger in Plato, see Jean-François Mattei, *L'Etranger et le simulacre* (Paris: P.U.F., 1983).

7. Only cursory allusions will be given here: to the bond of Eros with philia in the Greeks; to the role of the Fiancée and the Seducer in Kierkegaard; to the noetic function of the Couple according to Pierre Klossowski, *Les lois de l'hospitalité* (Paris: Gallimard, 1989); to the constitution of the woman-philosopher according to Michèle Le Doeuff, *Hipparchia's Choice,* trans. Trista Selous (Oxford: Basil Blackwell, 1991); and to the new persona of the Friend in Blanchot.

8. On this complex device, cf. Thomas de Quincey, "The Last Days of Immanuel Kant," in David Masson, ed., *Collected Writings,* vol. 4, pp. 340–41 (Edinburgh: Adam and Charles Black, 1890). [*Translators' note:* the passage reads, ". . . for fear of obstructing the circulation of the blood, he never would wear garters; yet, as he found it difficult to keep up his stockings without them, he had invented for himself a most elaborate substitute, which I will describe. In a little pocket, somewhat smaller than a watch-pocket, but occupying pretty nearly the same situation as a watch pocket on each thigh, there was placed a small box, something like a watch-case, but smaller; into this box was introduced a watch-spring in a wheel, round about which wheel was wound an elastic cord, for regulating the force of which there was a separate contrivance. To the two ends of this cord were attached hooks, which hooks were carried through a small aperture in the pockets, and so, passing down the inner and outer side of the thigh, caught hold of two loops which were fixed on the off side and the near side of each stocking."]

9. Søren Kierkegaard, "Fear and Trembling," in *Kierkegaard's Writings,* ed. and trans. Howard V. Hong and Edna H. Hong (Princeton: Princeton University Press, 1983), vol. 6, p. 49.

10. François Jullien, *Procès ou création* (Paris: Seuil, 1989), pp. 18, 117.

11. Nietzsche, *Musarion-Ausgabe* (n.p., n.d.), vol. 16, p. 35. Nietzsche often invokes a philosophical taste and derives "sage" from *sapere* (sapiens, the wine taster, sisyphos, the man of extremely "keen" taste). *Philosophy in the Tragic Age of the Greeks,* trans. Marianne Cowan (Chicago: Henry Regnery, 1962), p. 43.

12. Cf. Emile Bréhier, "La notion de problème en philosophie," in *Etudes de philosophie antique* (Paris: P.U.F., 1955).

13. Friedrich Nietzsche, *On the Genealogy of Morals,* trans. W. Kaufmann and R. J. Hollingdale (New York: Random House, 1967) vol. 1, 6.

4: Geophilosophy

1. These problems have been renewed profoundly by Marcel Detienne. On the opposition of the founding Stranger and the Autochthon, on the complex mixtures between these two poles, and on Erechtheus, see "Qu'estce que'un site?" in Marcel Detienne, ed., *Tracés de fondation* (Leuven: Peters, n.d.). Cf. also Giulia Sissa and Marcel Detienne, *La vie quotidienne des dieux grecs* (Paris: Hachette, 1989). For Erechtheus, see chap. 14; and for the difference between the two polytheisms, chap. 10.

2. V. Gordon Childe, *The Pre-History of European Society* (Harmondsworth: Penguin, 1958).

3. Jean-Pierre Faye, *La raison narrative: Langages totalitaires* (Paris: Balland, 1990), pp. 15–18. Cf. Clémence Ramnoux, in *Histoire de la philosophie* (Paris: Gallimard, n.d.), vol. 1, pp. 408–9: pre-Socratic philosophy is born and expands "on the edge of the Hellenic area as defined by colonization toward the end of the seventh and the beginning of the sixth century, and where, precisely through commerce and war, the Greeks confront the kingdoms and empires of the East"; it then wins over "the extreme west, the colonies of Sicily and Italy, thanks to migrations provoked by Iranian invasions and political revolutions." Friedrich Nietzsche, *Naissance de la philosophie* (Paris: Gallimard, 1969), p. 131: "Think of the philosopher as an émigré who has arrived among the Greeks; this is how it is for these pre-Platonists. In a way they are disorientated strangers."

4. On this pure sociability, "before and beyond any particular content," and democracy and conversation, cf. Georges Simmel, *Sociologie et épistémologie* (Paris: P.U.F., 1981), chap. 3.

5. Today, by freeing themselves from Hegelian or Heideggerian stereotypes, certain authors are taking up the specifically philosophical question on new foundations: on a Jewish philosophy, see the works by Lévinas and those around him, *Les cahiers de la nuit surveillée* 3 (1984); on an Islamic philosophy, according to the works of Corbin, see Christian Jambet, *La logique des Orientaux: Henry Corbin et la science des formes* (Paris: Seuil, 1983), and Guy Lardreau, *Discours philosophique et discours spirituel* (Paris: Seuil, 1985); on a Hindu philosophy, according to Masson-Oursel, see the approach of Roger-Pol Droit, *L'oubli de l'Inde: une amnésie philosophique* (Paris: P.U.F., 1989); on a Chinese philosophy, see the studies of François Cheng, *Vide et plein* (Paris: Seuil, 1991), and François Jullien, *Procès ou création* (Paris: Seuil, 1989); and on a Japanese philosophy, see René de Ceccaty and Ryôji Nakamura, *Mille ans de littérature japonaise,* and the translation with commentary of the monk Dôgen (Paris: La Différence, 1980).

6. Cf. Jean Beaufret: "The source is everywhere, undetermined, Chinese

as well as Arab and Indian . . . But then there is the Greek episode, the Greeks having had the strange privilege to call the source being" (*Ethernité* I [1985]).

7. Friedrich Nietzsche, "On the Uses and Disadvantages of History for Life," in *Untimely Meditations,* trans. R. J. Hollingdale (Cambridge: Cambridge University Press, 1983), I, pp. 63–64. On the philosopher-comet and the "environment" he finds in Greece, see Nietzsche, *Philosophy in the Tragic Age of the Greeks,* trans. Marianne Cowan (Chicago: Henry Regnery, 1962), pp. 33–34.

8. See Etienne Balazs, *Chinese Civilization and Bureaucracy,* trans. H. M. Wright (New Haven: Yale University Press, 1964), chap. 8.

9. Karl Marx, *Capital* (London: Lawrence and Wishart, 1972), vol. 3, part 3, chap. 15, p. 250: "Capitalist production seeks continually to overcome these immanent barriers, but overcomes them only by means which again place these barriers in its way and on a more formidable scale. The real barrier of capitalist production is capital itself."

10. Edmund Husserl, *The Crisis of European Sciences and Transcendental Phenomenology,* trans. D. Carr (Evanston, Ill.: Northwestern University Press, 1970). Cf. Droit's commentaries, *L'oubli de l'Inde,* pp. 203–4.

11. F. Braudel, *Capitalism and Material Life, 1400–1800,* trans. Miriam Kochan (New York: Harper and Row, 1967).

12. On these types of utopia, see Ernst Bloch, *Le principe espérance* (Paris: Gallimard, 1982), vol. 2. See also the commentaries on the relationship of Fourier's utopia with movement in René Schérer, *Pari sur l'impossible* (Paris: Presses universitaire de Vincennes, 1989).

13. Immanuel Kant, *The Contest of Faculties,* trans. Mary J. Gregor, (Lincoln: University of Nebraska Press, 1992), part 2, 6, pp. 153–57. The full importance of this text has been rediscovered today through the very different commentaries of Foucault, Habermas, and Lyotard.

14. Hölderlin: the Greeks possess the great panic Plane, which they share with the East, but they have to acquire the concept of Western organic composition; "with us, it is the other way round" (letter to Bölhendorf, 4 December 1801, with commentary by Jean Beaufret, in Friedrich Hölderlin, *Remarques sur Oedipe* (Paris: 10–18, 1965), pp. 8–11. See also Philippe Lacoue-Labarthe, *L'imitation des modernes* (Paris: Galilée, 1986). Even Renan's celebrated text on the Greek "miracle" has an analogous complex movement: what the Greeks possessed by nature we can rediscover only through reflection, by confronting a fundamental forgetfulness and worldweariness; we are no longer Greeks, we are Bretons: Ernest Renan, *Souvenirs d'enfance et de jeunesse* (Paris: Gallimard, 1983).

15. We refer to the first lines of the preface to the first edition of the *Critique of Pure Reason,* trans. N. Kemp-Smith (London: Macmillan,

1929), pp. 7–8: "The battle-field of these endless controversies is called metaphysics . . . Her government, under the administration of the *dogmatists,* was at first *despotic.* But inasmuch as the legislation still bore traces of the ancient *barbarism,* her empire gradually through intestine wars gave way to complete *anarchy*; and the *sceptics,* a species of *nomads,* despising all settled modes of life, broke up from time to time all civil society. Happily they were few in number, and were unable to prevent its being established ever anew, although on no uniform and self-consistent *plan.*" On the *island of foundation* we refer to the great text at the beginning of chap. 3 of the "Analytic of Principles" (p. 257). The critiques include not only a "history" but above all a "geography" of Reason, according to which a "field," a "territory" (*territorium*), and a "realm" (*ditio*) of the concept are distinguished (Immanuel Kant, *Critique of Judgement,* trans. J. H. Bernard [New York: Macmillan, 1951], introduction, sec. 2, "Of the Realm of Philosophy in General"). Jean-Clet Martin in his forthcoming *Variations* has produced a fine analysis of this geography of pure reason in Kant.

16. David Hume: "Two men, who pull the oars of a boat, do it by an agreement or convention, tho' they have never given promises to each other." *A Treatise of Human Nature* (Oxford: Clarendon Press, 1978), p. 490.

17. It is a "composite" feeling that Primo Levi describes in this way: shame that men could do this, shame that we have been unable to prevent it, shame at having survived, and shame at having been demeaned or diminished. See *The Drowned and the Saved,* trans. Raymond Rosenthal (London: Sphere Books, 1989), also on the "grey zone, with ill-defined outlines which both separate and join the two camps of masters and servants" (p. 27).

18. On the critique of "democratic public opinion," its American model, and the mystifications of human rights or of the State of international law, one of the strongest analyses is that of Michel Butel in *L'autre journal* 10 (March 1991): 21–25.

19. Charles Péguy, *Clio* (Paris: Gallimard, 1931), pp. 266–69.

20. Michel Foucault, *The Archaeology of Knowledge,* trans. A. M. Sheridan Smith (London: Tavistock, 1972), pp. 130–31.

5: Functives and Concepts

1. Ilya Prigogine and Isabelle Stengers, *Entre le temps et l'éternité* (Paris: Fayard, 1988), pp. 162–63. The authors take the example of the crystallization of a superfused liquid, a liquid at a temperature below its crystallization temperature: "In such a liquid, small germs of crystals form, but these germs appear and then dissolve without involving any consequences."

2. Georg Cantor, "Fondements d'une théorie générale des ensembles," in *Cahiers pour l'analyse* 10 (n.d.). From the beginning of the text Cantor invokes the Platonist Limit.

3. On the instituting of coordinates by Nicolas Oresme, intensive ordinates and their placing in relationship with extensive lines, cf. Pierre Duhem, *Le système du monde* (Paris: Hermann, 1913–59), vol. 7 (*La physique parisienne au XIVe siècle*), chap. 6. And, on the association of a "continuous spectrum and a discrete sequence," and Oresme's diagrams, see "La toile, le spectre, le pendule," in Gilles Châtelet's forthcoming *Les enjeux du mobile*.

4. G.W.F. Hegel, *Science de la logique* (Paris: Aubier, 1981), vol. 2, p. 277 (and on the operations of depotentialization and potentialization of the function according to Lagrange).

5. Pierre Vendryès, *Déterminisme et autonomie* (Paris: Armand Colin, n.d.). It is not the mathematization of biology that is of interest in the works of Vendryès but a homogenization of the mathematical and biological function.

6. On the meaning taken by the word *figure* (or *image, Bild*) in a theory of functions, see Vuillemin's analysis concerning Riemann: in the projection of a complex function, the figure "brings into view the course of the function and its different affections"; it "makes the functional correspondence" of the variable and the function "immediately visible." Jules Vuillemin, *La philosophie de l'algèbre* (Paris: P.U.F., 1962), pp. 320–26.

7. G. W. Leibniz, "D'une ligne issue de lignes" and "Nouvelle application du calcul," both in *Oeuvre mathématique de Leibniz autre que le calcul infinitesimal*, trans. Jean Peyroux (Paris: Blanchard, 1986). These texts are considered to be the foundations of the theory of functions.

8. Having described the "intimate mixture" of different types of trajectory in every region of the phase space of a system with weak stability, Prigogine and Stengers conclude: "We may think of a familiar situation, that of the numbers on the axis where each rational number is surrounded by irrational numbers, and each irrational number is surrounded by rational numbers. Equally, we may think of the way in which Anaxagoras [shows how] every thing contains in all its parts, even the smallest, an infinite multiplicity of qualitatively different seeds intimately mixed together." Ilya Prigogine and Isabelle Stengers, *La nouvelle alliance* (Paris: Gallimard, 1979), p. 241. [*Translators' note:* the English version of this book, *Order out of Chaos* (London: HarperCollins, 1985), differs considerably from the original French, but see p. 264 of the English version.]

9. The theory of two kinds of "multiplicity" is present in Bergson from *Time and Free Will,* trans. F. L. Pogson (New York: Macmillan, 1910), chap. 2: multiplicities of consciousness are defined by "fusion" and by "penetration" terms that are equally found in Husserl from *The Philosophy of*

Arithmetic. The resemblance between the two authors is, in this respect, extremely close. Bergson will always define the object of science by mixtures of space-time, and its principal action by the tendency to take time as an "independent variable," whereas, at the other pole, duration passes through every variation.

10. Gilles-Gaston Granger, *Essai d'une philosophie du style* (Paris: Odile Jacob, 1988), pp. 10–11, 102–5.

11. Cf. the great texts of Evariste Galois on mathematical enunciation: André Dalmas, *Evariste Galois, revolutionnaire et geometre* (Paris: Nouveau Commerce, 1982), pp. 117–32.

12. Jacques Monod, *Chance and Necessity,* trans. Austryn Wainhouse (Glasgow: Collins/Fount Paperbacks, 1977), p. 78: "Allosteric interactions are indirect, proceeding exclusively from the protein's discriminatory properties of stereospecific recognition, in the two (or more) states accessible to it." A process of molecular recognition may introduce very different mechanisms, thresholds, sites, and observers, as in the recognition of male-female in plants.

13. Bertrand Russell, "The Relation of Sense-Data to Physics," in *Mysticism and Logic* (London: Longmans, Green, 1918).

14. Throughout his work, Bergson opposes the scientific observer to the philosophical persona who "passes" through duration. In particular, he tries to show that the former presupposes the latter, not only in Newtonian physics (*Time and Free Will,* trans. F. L. Pogson [New York: Macmillan, 1910], chap. 3) but in relativity (*Duration and Simultaneity,* trans. Leon Jacobson [Indianapolis: Bobbs-Merrill, 1965]).

6: Prospects and Concepts

1. Cf. Bertrand Russell, *The Principles of Mathematics* (London: Unwin, 1903), especially appendix A; and Gottlob Frege, *The Foundations of Arithmetic* (Oxford: Basil Blackwell, 1953), 48 and 54, and *Translations from the Philosophical Writings of Gottlob Frege,* trans. and ed. Peter Geach and Max Black (Oxford: Basil Blackwell, 1979), especially the papers "Function and Concept," "On Concept and Object," and, for the critique of the variable, "What Is a Function?" See also Claude Imbert's commentaries on these works in the French translations of Frege: *Les fondements de l'arithmétique* (Paris: Seuil, 1970) and *Ecrits logiques et philosophique* (Paris: Seuil, 1971). See also Philippe de Rouilhan, *Frege, les paradoxes de la représentation* (Paris: Minuit, 1988).

2. Oswald Ducrot has criticized the self-referential character attributed to performative statements (what one does by saying it: I swear, I promise, I order): *Dire et ne pas dire* (Paris: Hermann, 1980), pp. 72f.

3. On projection and Gödel's method, see E. Nagel and J. R. Newman, *Gödel's Proof* (London: Routledge, 1959).

4. On Frege's conception of the interrogative proposition, see Gottlob Frege, *Logical Investigations,* trans. P. T. Geach and R. H. Stoothoff (Oxford: Basil Blackwell, 1977), and also for the three elements: grasping thought, or the act of thinking; recognition of the truth of a thought, or judgment; the expression of judgment, or affirmation. See also Russell, *The Principles of Mathematics,* 477.

5. For example, between true and false (1 and o), degrees of truth are introduced that are not probabilities but produce a kind of fractalization of the peaks of truth and the troughs of falsity, so that the fuzzy sets become numerical again, but through a fractional number between o and 1. However, this is on condition that the fuzzy set is the subset of a normal set, referring to a regular function. See Arnold Kaufmann, *Introduction to the Theory of Fuzzy Subsets,* trans. D. L. Swanson (New York: Academic Press, 1975), and Pascal Engel, who devotes a chapter to the "vague" in *Le norme du vrai: philosophie de la logique* (Paris: Gallimard, 1989).

6. On the three transcendences that appear within the field of immanence, the primordial, the intersubjective, and the objective, see Edmund Husserl, *Cartesian Meditations,* trans. Dorion Cairns (The Hague: Martinus Nijhoff, 1960), especially 55–56. On the *Urdoxa,* see *Ideas, General Introduction to Pure Phenomenology,* trans. W. R. Boyce Gibson (New York: Humanities Press, 1969), especially 103–4; and E. Husserl, *Experience and Judgement: Investigations in a Genealogy of Logic,* trans. James S. Churchill and Karl Ameriks (Evanston, Ill.: Northwestern University Press, 1973).

7. G.-G. Granger, *Pour la connaissance philosophique* (Paris: Odile Jacob, 1988), chaps. 6 and 7. Knowledge of the philosophical concept is reduced to reference to the lived, inasmuch as the latter constitutes it as "virtual totality": this implies a transcendental subject, and Granger seems to give "virtual" no other meaning than the Kantian one of a whole of possible experience (pp. 174–75). It is noticeable that Granger gives a hypothetical role to "fuzzy concepts" in the transition from scientific to philosophical concepts.

8. On abstract thought and popular judgment, see the short text by Hegel entitled "Qui pense abstrait?" in *Sämtliche Werke,* vol. 20, pp. 445–50.

9. Marcel Detienne shows that philosophers lay claim to a knowledge that is distinct from the old wisdom and to an opinion that is distinct from that of the sophists. Marcel Detienne, *Les maîtres de vérité dans la Grèce archaïque* (Paris: Maspero, 1973), chap. 6, pp. 131ff.

10. See Heidegger's celebrated analysis, and Beaufret's, in Jean Beaufret, ed., *Le poème* (Paris: P.U.F., 1986), pp. 31–34.

11. Alain Badiou, *L'être et l'événement* (Paris: Seuil, 1988), and *Manifeste*

pour la philosophie (Paris: Seuil, 1989). Badiou's theory is very complex; we fear we may have oversimplified it.

12. See Alfred North Whitehead, *Process and Reality* (New York: Free Press, 1979), pp. 22–26.

13. Paul Klee, *On Modern Art,* trans. Paul Findlay (London and Boston: Faber, 1966), p. 45.

14. Science feels the need not only to order chaos but to see it, touch it, and produce it: cf. James Gleick, *Chaos: Making a New Science* (London: Sphere, 1988). Gilles Châtelet in his forthcoming *Les enjeux du mobile* shows how mathematics and physics attempt to retain something of a sphere of the virtual.

15. Péguy, *Clio,* pp. 230, 265. Maurice Blanchot, *The Space of Literature,* trans. Ann Smock (Lincoln: University of Nebraska Press, 1989), pp. 90, 122, 126.

16. James Gleick, *Chaos,* p. 186.

17. On the meanwhile [*l'entre-temps*], we refer to a very intense article by B. Groethuysen, "De quelques aspects du temps," *Recherches philosophiques* 5 (1935–36): "All events are, so to speak, in the time where nothing is happening." All of Lernet-Holonia's novelistic work takes place in meanwhiles.

18. Joe Bousquet, *Les Capitales* (Paris: Le Cercle du livre, 1955), p. 103.

19. Stéphane Mallarmé, "Mimique," *Oeuvres complètes* (Paris: La Pléiade, Gallimard, 1945), p. 310.

7: Percept, Affect, and Concept

1. Edith Wharton, *Les metteurs en scène* (Paris: 10–18, 1986), p. 263. It concerns an academic and worldly painter who gives up painting after seeing a little picture by one his unrecognized contemporaries: "And me, I have not created any of my works, I have simply adopted them."

2. Joachim Gasquet, *Cézanne: A Memoir with Conversations,* trans. Christopher Pemberton (London: Thames and Hudson, 1991), p. 164.

3. See François Cheng, *Vide et plein* (Paris: Seuil, 1979), p. 63 (citation of the painter Huang Pin-Hung).

4. Antonin Artaud, "Van Gogh: The Man Suicided by Society," in Jack Hirschman, ed., *Artaud Anthology* (San Francisco: City Lights Books, 1965), pp. 156, 160 (translation modified): "As a painter, and nothing else but a painter, Van Gogh adopted the methods of pure painting and never went beyond them. . . . The marvelous thing is that this painter who was only a painter . . . among all the existing painters, is [also] the one who makes us forget that we are dealing with painting" (pp. 154–56).

5. José Gil devotes a chapter to the procedure by which Pessoa extracts

the percept on the basis of lived perceptions, particularly in "L'ode maritime." *Fernando Pessoa ou la métaphysique des sensations* (Paris: La Différence, 1988), chap. 2.

6. Gasquet, *Cézanne,* p. 154. See Erwin Straus, *Du sens des sens* (Paris: Millon, n.d.), p. 519: "The great landscapes have a wholly visionary characteristic. Vision is what of the invisible becomes visible . . . The landscape is invisible because the more we conquer it, the more we lose ourselves in it. To reach the landscape we must sacrifice as much as we can all temporal, spatial, objective determination; but this abandon does not only attain the objective, it affects us ourselves to the same extent. In the landscape we cease to be historical beings, that is to say, beings who can themselves be objectified. We do not have any memory for the landscape, we no longer have any memory for ourselves in the landscape. We dream in daylight with open eyes. We are hidden to the objective world, but also to ourselves. This is feeling."

7. Roberto Rossellini, *Le cinéma révélé* (Paris: Etoile-Cahiers du cinéma, 1984), pp. 80–82.

8. In the second chapter of *The Two Sources of Morality and Religion,* trans. T. Ashley Audra and Cloudesley Brereton with the assistance of W. Horsfall Carter (New York: Henry Holt, 1935), Bergson analyzes fabulation as a visionary faculty very different from the imagination and that consists in creating gods and giants, "semi-personal powers or effective presences." It is exercised first of all in religions, but it is freely developed in art and literature.

9. Virginia Woolf, *The Diary of Virginia Woolf,* ed. Anne Olivier Bell (London: Hogarth Press, 1980), vol. 3, pp. 209, 210.

10. Antonin Artaud, *The Theatre and Its Double,* trans. Mary Caroline Richards (New York: Grove Press, 1958), p. 134.

11. Jean-Marie Gustave Le Clézio, *HAI* (Paris: Flammarion, 1991), p. 7 ("I am an Indian"—although I do not know how to cultivate corn or make a dugout). In a famous text, Michaux spoke of the "health" peculiar to art: postface to "Mes propriétés"–in Henri Michaux, *La nuit remue* (Paris: Gallimard, 1935), p. 193.

12. André Dhôtel, *Terres de mémoire,* (Paris: Presses Universitaires de France, 1979), pp. 225–26.

13. Emile Bergson, *The Creative Mind,* trans. Mabelle L. Andison (Westport, Conn.: Greenwood Press, 1946), pp. 59–60.

14. These three questions frequently recur in Proust, especially in "Time Regained" in *Remembrance of Things Past,* trans. C. K. Scott Moncrieff and Terence Kilmartin; and by Andreas Mayor (London: Chatto and Windus, 1982), vol. 3, pp. 931–32 (on life, vision, and art as the creation of universes).

15. Malcolm Lowry, *Under the Volcano* (Harmondsworth: Penguin, 1963), p. 183.

16. Osip Mandelstam, *The Noise of Time: The Prose of Osip Mandelstam,* trans. with critical essays by Clarence Brown (San Francisco: North Point Press, 1986), pp. 109–10.

17. Mikel Dufrenne, in *Phénoménologie de l'expérience esthétique* (Paris: P.U.F., 1953), produced a kind of analytic of perceptual and affective a priori, which founded sensation as a relationship of the body and the world. He stayed close to Erwin Straus. But is there a being of sensation that manifests itself in the flesh? Maurice Merleau-Ponty followed this path in *The Visible and the Invisible,* trans. A. Lingis (Evanston, Ill.: Northwestern University Press, 1969). Dufrenne emphasized a number of reservations concerning such an ontology of the flesh (see *L'oeil et l'oreille* [Montreal: Hexagone, 1987]). Recently, Didier Franck has again taken up Merleau-Ponty's theme by showing the decisive importance of the flesh in Heidegger and already in Husserl (see *Heidegger et le problème de l'espace* [Paris: Minuit, 1986] and *Chair et corps* [Paris: Minuit, 1981]). This whole problem is at the center of a phenomenology of art. Perhaps Michel Foucault's still-unpublished book *Les aveux de la chair* will teach us about the most general origins of the notion of the flesh and its significance in the Church Fathers.

18. As Georges Didi-Huberman demonstrates, the flesh gives rise to a "doubt": it is too close to chaos. Hence the necessity of a complementarity between the "pink" [*incarnat*] and the "section" [*pan*], the essential theme of *La peinture incarnée* (Paris: Minuit, 1985), which is taken up again and developed in *Devant l'image* (Paris: Minuit, 1990).

19. Vincent Van Gogh, letter no. 520 to Theo, 11 August 1888, in *The Complete Letters of Vincent Van Gogh* (Greenwich, Conn.: New York Graphic Society, 1958), vol. 3. Broken tones and their relationship with the area of plain, uniform color are a frequent theme of the correspondence. Similarly for Gauguin; see letter to Schuffenecker, 8 October 1888, in *Lettres* (Paris: Grasset, 1946), p. 140: "I have done a self-portrait for Vincent . . . I think it is one of my best: absolutely incomprehensible (for example) it is so abstract . . . its drawing is completely special, complete abstraction . . . The color is a color far from nature; imagine a vague memory of pottery buckled by great heat. All the reds, the violets, scored by the fire's blaze like a furnace glowing to the eyes, seat of the struggles of the painter's thought. All on a chrome ground sprinkled with childish bunches of flowers. Room of pure young girl." This is the idea of the "arbitrary colorist" according to Van Gogh.

20. Cf. *Artstudio* (n.d.), no.16, "Monochromes" (Geneviève Monnier and Denys Riout on Klein, and Pierre Sterckx on the "current avatars of monochrome").

21. Wilhelm Worringer, *Form in Gothic* (London: Putnam's and Sons, 1927).

22. Piet Mondrian, "Réalité naturelle et réalité abstraite," in Michel Seuphor, *Piet Mondrian, sa vie, sa oeuvre* (Paris: Flammarion, n.d.), on the room and its unfolding. Michel Butor has analyzed this unfolding of the room into squares or rectangles, and the opening onto an interior square, empty and white like the "promise of a future room." Michel Butor, "Le carré et son habitant," *Répertoire III* (Paris: Minuit, 1992), pp. 307–9, 314–15.

23. It seems to us that Lorenz's mistake is wanting to explain the territory by an evolution of functions: Konrad Lorenz, *On Aggression,* trans. Marjorie Kerr Wilson (New York: Harcourt, Brace and World, 1966).

24. Alan John Marshall, *Bower Birds* (Oxford: Clarendon Press, 1954); and E. T. Gilliard, *Birds of Paradise and Bower Birds* (London: Weidenfeld, 1969).

25. See Jakob von Uexküll's masterpiece, *Mondes animaux et monde humain, Théorie de la signification* (Paris: Gonthier, 1965), pp. 137–42: "counterpoint, motif of development, and morphogenesis."

26. Henry van de Velde, *Déblaiement d'art* (Brussels: Archives architecture moderne, 1979), p. 20.

27. On all these points, the analysis of enframing forms, and of the town-cosmos (the example of Lausanne), see Bernard Cache's forthcoming *L'ameublement du territoire.*

28. Pascal Bonitzer formed the concept of deframing [*décadrage*] in order to highlight new relationships between the planes in cinema (*Cahiers du cinéma* 284 [January 1978]): "disjointed, crushed or fragmented" planes, thanks to which cinema becomes an art by getting free from the commonest emotions, which were in danger of preventing its aesthetic development, and by producing new affects. See Pascal Bonitzer, *Le champ aveugle: essais sur le cinéma* (Paris: Gallimard–Cahiers du Cinéma, 1982): "system of the emotions."

29. Mikhail Bakhtine, *Esthétique et théorie du roman,* trans. Daria Olivier (Paris: Gallimard, 1978).

30. Pierre Boulez, especially *Orientations,* trans. Martin Cooper (London: Faber, 1986), and *Boulez on Music Today,* trans. S. Bradshaw and Richard Rodney Bennett (London: Faber, 1971). The extension of the series into durations, intensities, and timbres is not an act of closure but, on the contrary, an opening of what is closed in the series of pitches [*hauteurs*].

31. Xavier de Langlais, *La technique de la peinture à l'huile* (Paris: Flammarion, 1988); Johann Wolfgang von Goethe, *Theory of Colors* (Cambridge, Mass.: MIT Press, 1970), 902–9.

32. See Christian Bonnefoi, "Interview et comment par Yves-Alain Bois," *Macula* (n.d.), 5–6.

33. Hubert Damisch, *Le fenêtre jaune cadmium; ou Les dessous de la*

peinture (Paris: Seuil, 1984), pp. 275–305 (and p. 80, on the thickness of the plane in Pollock). Damisch has insisted more than other writers on art-as-thought and painting-as-thought, such as Dubuffet in particular sought to institute. Mallarmé made the book's "thickness" a dimension distinct from its depth; see Jacques Schérer, *Le Livre de Mallarmé* (Paris: Gallimard, 1978), p. 55. Boulez takes up this theme on his own account for music (*Orientations*).

Conclusion: From Chaos to the Brain

1. Immanuel Kant, *Critique of Pure Reason,* trans. N. Kemp-Smith (London: Macmillan, 1929), Transcendental Analytic, "The Synthesis of Reproduction in Imagination."

2. On Cézanne and chaos, see Gasquet, *Cézanne*; on Klee and chaos, see Paul Klee, "Note on the Gray Point," in *Théorie de l'art moderne* (Paris: Gonthier, 1963). See also the analyses of Henri Maldiney, *Regard Parole Espace* (Paris: L'Age d'homme, 1973), pp. 150–51, 183–85.

3. Galois, in Dalmas, *Evariste Galois,* pp. 121, 130.

4. Lawrence, "Chaos in Poetry," in D. H. Lawrence, *Selected Literary Criticism,* ed. A. Beal (London: Heinemann, 1955).

5. Georges Didi-Huberman, *La peinture incarnée* (Paris: Minuit, 1985), pp. 120–23, on the flesh and chaos.

6. Michel Serres, *Le système de Leibniz* (Paris: P.U.F., 1990), vol.1, p. 111 (and pp. 120–23, on the succession of filters).

7. On strange attractors, independent variables, and "routes toward chaos," see Prigogine and Stengers, *Entre le temps et l'éternité,* chap. 4, and James Gleick, *Chaos.*

8. See Martial Guéroult, *L'évolution et la structure de la Doctrine de la science chez Fichte* (Paris: Belles Lettres, 1982), vol. 1, p. 174.

9. Jean-Clet Martin's forthcoming *Variations.*

10. Erwin Straus, *Du sens des sens,* part 3.

11. Raymond Ruyer, *Néo-finalisme* (Paris: P.U.F., 1952). Throughout his work Ruyer has directed a double critique against mechanism and dynamism (Gestalt), which differs from the critique made by phenomenology.

12. David Hume defines imagination by this passive contemplation-contraction: *A Treatise of Human Nature* (Oxford: Clarendon Press, 1978), book 1, part 3, 14.

13. Plotinus's great text on contemplations is at the beginning of *Enneades* 3.8. The empiricists, from Hume to Butler to Whitehead, will take up the theme by inclining it toward substance; hence their neo-Platonism.

14. Burns, *The Uncertain Nervous System* (London: Edward Arnold,

n.d.). See also Steven Rose, *The Conscious Brain* (New York: Knopf, 1975): "The nervous system is uncertain, probabilistic, and so interesting."

15. Immanuel Kant, *Critique of Judgement,* trans. J. H. Bernard (New York: Macmillan, 1951), 62.

16. François Laruelle proposes a comprehension of nonphilosophy as the "real (of) science," beyond the object of knowledge: *Philosophie et nonphilosophie* (Liege: Mardaga, 1989). But we do not see why this real of science is not nonscience as well.

Index

Absolute, of a concept, 21
Abstract art: forces creating, 181; nature of, 198
Accumulation, point of, in concepts, 20
Actual, the, 113
Adorno, negative dialectic and, 99
Aesthetic figures, *see* Figures
Affects, 66; art extracts, 24; beyond affection, 173; concepts and percepts and, 163–99; *see also* Sensation
Agon as rule of a society of friends, 9
Aleph 0, 120
Allosteric enzymes, observers and observing and, 130, 227n12
America: philosophy in, 143; revolution in, turned out so badly, 100
Anaximander, 44; limit understood by, 120
Animal, the, in art, 184–85
Animism, biology and, 130
Another person, self relating to, 16
Antipathetic characters, as conceptual personae, 63
Aporia of Platonists, 148, 149
Aristocracy, future and, 108
Aristotle, opposable opinion in, 79
Art: abstract, nature of, 198; extracts affects, 24; artist and drugs in, 165; brain and, 211–12; brings varieties from chaos, 202; chaos and, 202–7; color in, 181; planes of composition and immanence and, 66; composition in, 191–97; as composition of chaos, 204; conceptual, nature of, 198; deterritorialization in, 181; figures in, 66, 187, 196; finite in, 197–98; framing in, 198; infinite in,

197–98; the infinite in, 181; as language of sensation, 176; material in, 191–97; memory in, 167; nonart and, 218; extracts percepts, 24; philosophy and, of portrait, 55–57; progress in, 193; relationship of, to philosophy and science, 5; resemblance in, 166; sensation in, 179; thought in, 197–99; vibrations in, 168
Artaud, 109; on immanence, 49; thought and, 55
Association of ideas, meaning of, 201
Aternal, Péguy's term of, defined, 111–12, 157
Atheism: in art, 194; figures versus concepts and thought in, 92
Athleticism in literature, 172
Attractors, strange or chaotic, 206, 233n7
Autobiography in novels, 170
Autochthon, the: deterritorialization and, 86, 98; for the Greeks, 101–2; and the stranger, 110, 223n1

Bacon, Francis (1909–), infinite in, 181
Badiou, Alain, on concepts and functions in, 151
Bakhtin, theory of the novel of, 188
Balzac, H. de, 171
Barnes, Hazel, 3n
Bartók, Béla, framing in, 191
Beast, the, as literary character, 174
Beaumarchais, P. A. C. de, 106
Beckett, Samuel, 174
Becoming, 158; concepts and, 18, 20; diagnosis of, 113; as always double,

migrant as, 67, 98; music and aesthetic figures and, 65; native as, 67; nature of, 61–83; Nietzsche and, 65; noumenon and, 65; in the novel, 65; objectality and, 3, 4n; in paintings, 65; pathic features and, 70; phenomenon and, 65; Philalethes as friend as, 3; in philosophy, 2–12; plaintiff as, 72; plane of composition and, 65; proper names as, 24; psychosocial types and, 67, 70; reason and, 77; relational features and, 70; repulsive characters as, 63; reterritorialization and, 67–68; rival as, 4; schoolman as, 62; in sculpture, 65; as sensibilia, 131; social fields and, 68; societies of friends or equals and, 4; Socrates as, for Platonism, 63, 65; Stammerer as, 69; stranger as, 69; as subject of philosophy, 64; surfer as, 71; taste and, 77, 78, 133; territory and, 67–68; Theophilus as friend as, 3; thing as, 4; as thinkers, 69; third party as, 4; thought and, 4–5; Thought-Being and, 65; thought-events and, 70; transient as, 67; universe and, 65, 177; *see also* Concept(s)

Concettism, Italy and Spain and, 103
Condensation, point of, in concepts, 20
Consistency: chaos undoes, 42; of concepts, 19, 22, 126, 137; element presented by philosophy, 77
Constructivism: concepts and, 7; conceptual personae and, 75; disqualifies all discussion, 82; philosophy as a, 35–36; relative and absolute of concepts and, 22; three activities constituting, 81; two different qualitative aspects of, 35–36
Contemplation: Eidetic era and, 47;

as illusion, 49; objectality of, as figure of philosophy, 92; philosophy not, 6; sensation as pure, 212; as universal, 15
Continuum in set theory, 120
Conversation: discussion, 28–29, 79; and thought, 140
Coordination, plane of, *see* Reference, plane of
Cosmos, 180, 189; *see also* Universe
Counterpoint: framing in art and, 187; in literature, 188
Creating, member of philosophical trinity, 77
Critical era, universal of reflection and, 47

Damisch, Hubert, material in composition and, 193, 195
Danger, new meaning of, with pure immanence, 42
Death, assimilated to, 161
Debussy, Claude, material in composition and, 195
Deleuze, Gilles, 2n
Delirium, plane of immanence and, 53
Democracy: capitalism and, 97, 98, 106; development of, 97; future and, 110; majorities and, 108; realization and new models of, 106; social, 107
Descartes, Réne: cogito of, 24–27; concept of self of, 24; concepts of, 24, 29–32; error and, 52; *see also* Cogito (Cartesian)
Deterritorialization: in art, 181; conceptual persona and, 67–68; earth and, 85; states and cities and, 86; of thought, 69, 70; when absolute, 88; *see also* Reterritorialization; Territory
Dhotel, André, characters of, 173